NEUROMETHODS

Series Editor
Wolfgang Walz
University of Saskatchewan,
Saskatoon, SK, Canada

For further volumes:
http://www.springer.com/series/7657

Neuromethods publishes cutting-edge methods and protocols in all areas of neuroscience as well as translational neurological and mental research. Each volume in the series offers tested laboratory protocols, step-by-step methods for reproducible lab experiments and addresses methodological controversies and pitfalls in order to aid neuroscientists in experimentation. *Neuromethods* focuses on traditional and emerging topics with wide-ranging implications to brain function, such as electrophysiology, neuroimaging, behavioral analysis, genomics, neurodegeneration, translational research and clinical trials. *Neuromethods* provides investigators and trainees with highly useful compendiums of key strategies and approaches for successful research in animal and human brain function including translational "bench to bedside" approaches to mental and neurological diseases.

Experimental and Translational Methods to Screen Drugs Effective Against Seizures and Epilepsy

Edited by

Divya Vohora

Neurobehavioral Pharmacology Laboratory, Department of Pharmacology,
School of Pharmaceutical Education and Research (SPER),
Jamia Hamdard, New Delhi, India

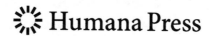 Humana Press

Editor
Divya Vohora
Neurobehavioral Pharmacology Laboratory
Department of Pharmacology
School of Pharmaceutical Education and Research (SPER)
Jamia Hamdard, New Delhi, India

ISSN 0893-2336   ISSN 1940-6045   (electronic)
Neuromethods
ISBN 978-1-0716-1256-9   ISBN 978-1-0716-1254-5   (eBook)
https://doi.org/10.1007/978-1-0716-1254-5

This Humana imprint is published by the registered company Springer Science+Business Media, LLC, part of Springer
Nature.
The registered company address is: 1 New York Plaza, New York, NY 10004, U.S.A.

Foreword

Epilepsies are an unpredictable group of diseases that can affect the lives of 1 in 26 people. It is an unpredictable group of ever-changing conditions that can manifest with seizures and is also associated with comorbidities in cognitive and psychiatric spheres as well as various degrees of brain damage. Recent work has identified that there are age- and gender-related differences in the expression and outcome of seizures. This is especially pertinent as some of the currently available treatment modalities do not help 30% of the people that suffer from epilepsy; even in cases where antiseizure drugs can control seizures, there may be a price to pay in terms of unacceptable side effects. With the recent advances in brain research and availability of innovative technologies, this is the time to promote epilepsy research. Effective research can change this potentially devastating group of diseases and provide a better future for people with epilepsy.

This *Neuromethods* volume titled *Experimental and Translational Models to Screen Drugs Effective Against Seizures and Epilepsy*, edited by Prof. Dr. Divya Vohora, provides detailed experimental and clinical approaches on seizure/epilepsy-related research. The volume includes contributions by outstanding researchers and covers the full spectrum of questions that need to be addressed in order to improve the care of people with epilepsy as a function of age, sex, and underlying etiology. Accordingly, it covers multiple experimental models and methods as all can be helpful in understanding the various manifestations of diseases associated with epilepsy. Each chapter has a specific outline, describing step by step the rationale of its selection, its value, and how to perform these studies.

This volume will be an indispensable tool for new researchers interested in choosing a model to study epilepsy and will provide valuable information to established investigators as well as industry partners looking to improve the care of people with epilepsy. This book is truly a global effort as shown by the breadth of investigators involved in this often-neglected field. Now is the time to move epilepsy-related research forward. This book is a valuable and effective step.

Solomon L. Moshé
Charles Frost Chair in Neurosurgery and Neurology
Professor of Neurology, Neuroscience & Pediatrics
Vice-Chair, Saul R Korey Department of Neurology, Bronx, NY, USA
Director, Isabelle Rapin Division of Child Neurology, Bronx, NY, USA
Director, Clinical Neurophysiology
Albert Einstein College of Medicine and Montefiore Medical Center, Bronx, NY, USA
Former President, International League Against Epilepsy (ILAE), Flower Mound, TX, USA
Former President, American Epilepsy Society (AES), Chicago, IL, USA

Preface to the Series

Experimental life sciences have two basic foundations: concepts and tools. The *Neuromethods* series focuses on the tools and techniques unique to the investigation of the nervous system and excitable cells. It will not, however, shortchange the concept side of things as care has been taken to integrate these tools within the context of the concepts and questions under investigation. In this way, the series is unique in that it not only collects protocols but also includes theoretical background information and critiques which led to the methods and their development. Thus, it gives the reader a better understanding of the origin of the techniques and their potential future development. The *Neuromethods* publishing program strikes a balance between recent and exciting developments like those concerning new animal models of disease, imaging, in vivo methods, and more established techniques, including, for example, immunocytochemistry and electrophysiological technologies. New trainees in neurosciences still need a sound footing in these older methods in order to apply a critical approach to their results.

Under the guidance of its founders, Alan Boulton and Glen Baker, the *Neuromethods* series has been a success since its first volume published through Humana Press in 1985. The series continues to flourish through many changes over the years. It is now published under the umbrella of Springer Protocols. While methods involving brain research have changed a lot since the series started, the publishing environment and technology have changed even more radically. *Neuromethods* has the distinct layout and style of the Springer Protocols program, designed specifically for readability and ease of reference in a laboratory setting.

The careful application of methods is potentially the most important step in the process of scientific inquiry. In the past, new methodologies led the way in developing new disciplines in the biological and medical sciences. For example, physiology emerged out of anatomy in the nineteenth century by harnessing new methods based on the newly discovered phenomenon of electricity. Nowadays, the relationships between disciplines and methods are more complex. Methods are now widely shared between disciplines and research areas. New developments in electronic publishing make it possible for scientists that encounter new methods to quickly find sources of information electronically. The design of individual volumes and chapters in this series takes this new access technology into account. Springer Protocols makes it possible to download single protocols separately. In addition, Springer makes its print-on-demand technology available globally. A print copy can therefore be acquired quickly and for a competitive price anywhere in the world.

Preface

More than 50 million people are currently suffering from epilepsy all over the world, and 2.4 million cases are diagnosed yearly. The seizures can be controlled up to a certain limit in some persons, whereas around 30% suffer from severe, uncontrolled, or refractory (also referred as intractable or pharmacoresistant) epilepsy. This unmet clinical need has mandated the development of new experimental models to screen novel drug molecules for seizures and epilepsy.

This volume discusses the detailed experimental and clinical approaches to study seizures, along with epilepsy. The methods covered are either widely employed traditionally used models or the recent approaches to screen antiseizure drugs including recent modifications of the traditional models. Though it was not possible to cover all the available models used in epilepsy research for this present volume given the highly heterogeneous nature of epilepsy with multiple seizure types and syndromes, the volume covers detailed protocols for ex vivo and in vitro experimental models and the very useful and widely employed in vivo (acute and chronic seizure) models used in epilepsy research. In addition, some of the methods to screen the effective leads for the treatment of refractory seizures or pharmacoresistant epilepsy and invertebrate/non-mammalian models are also included.

Each chapter incorporates the significance of the model/method, the rationale of its selection, its translational value, the materials required, adopted techniques, step-by-step methodology, as well as the optimum experimental conditions for the same. We hope that a detailed representation of each model/method by the experts and researchers in this area would improve the reproducibility, and would help the reader to work with similar conditions. Not only this, the tips and tricks involving troubleshooting, data analysis, and representation in addition to other important considerations are covered for anyone to work in this area.

This volume, with expertise from all the discussed areas, would hopefully prove to be a valuable tool to carry out and proceed in epilepsy research from scratch for both industry and the academia, which would further enhance the use of these techniques/methods in a systematic way for the field of experimental and translational epilepsy research further contributing to the community health.

New Delhi, India   *Divya Vohora*

Contents

Contributors

MD. JOYNAL ABEDIN • *Department of Bioengineering, Lehigh University, Bethlehem, PA, USA*

NIDHI BHARAL AGARWAL • *Centre for Translational and Clinical Research, School of Chemical and Life Sciences (SCLS), Jamia Hamdard, New Delhi, Delhi, India*

JYOTIRMOY BANERJEE • *Centre of Excellence for Epilepsy, A joint collaboration between All India Institute of Medical Sciences (AIIMS) and National Brain Research Centre (NBRC), New Delhi, India; Department of Biophysics, All India Institute of Medical Sciences, New Delhi, India*

MELISSA BARKER-HALISKI • *Department of Pharmacy, University of Washington, Seattle, WA, USA*

YEVGENY BERDICHEVSKY • *Department of Bioengineering, Lehigh University, Bethlehem, PA, USA; Department of Electrical Engineering, Lehigh University, Bethlehem, PA, USA*

EDWARD H. BERTRAM • *Department of Neurology, University of Virginia Health Sciences Center, Charlottesville, VA, USA*

P. SARAT CHANDRA • *Centre of Excellence for Epilepsy, A joint collaboration between All India Institute of Medical Sciences (AIIMS) and National Brain Research Centre (NBRC), New Delhi, Delhi, India; Department of Neurosurgery, All India Institute of Medical Sciences, New Delhi, Delhi, India*

BRANDON KAR MENG CHOO • *Neuropharmacology Research Strength, Jeffrey Cheah School of Medicine & Health Sciences, Monash University Malaysia, Bandar Sunway, Selangor, Malaysia*

VINCENT T. CUNLIFFE • *Department of Biomedical Science, University of Sheffield, Sheffield, UK*

SOUMIL DEY • *Centre of Excellence for Epilepsy, A joint collaboration between All India Institute of Medical Sciences (AIIMS) and National Brain Research Centre (NBRC), New Delhi, Delhi, India; Department of Neurosurgery, All India Institute of Medical Sciences, New Delhi, Delhi, India*

APARNA BANERJEE DIXIT • *Centre of Excellence for Epilepsy, A joint collaboration between All India Institute of Medical Sciences (AIIMS) and National Brain Research Centre (NBRC), New Delhi, Delhi, India; Dr. B R Ambedkar Centre for Biomedical Research, University of Delhi, New Delhi, Delhi, India*

SHABNAM GHIASVAND • *Department of Bioengineering, Lehigh University, Bethlehem, PA, USA*

SHRESHTA JAIN • *Neurobehavioral Pharmacology Laboratory, Department of Pharmacology, School of Pharmaceutical Education and Research (SPER), Jamia Hamdard, New Delhi, Delhi, India*

ALISTAIR JONES • *Institute of Translational Medicine, University of Liverpool, Liverpool, UK*

SHILPA D. KADAM • *Neuroscience Laboratory, Hugo Moser Research Institute at Kennedy Krieger, Baltimore, MD, USA; Department of Neurology, Johns Hopkins University School of Medicine, Baltimore, MD, USA*

RAZIA KHANAM • *Department of Biomedical Sciences, Gulf Medical University, Ajman, UAE*

JING LIU • *Department of Neurological Surgery, University of California, San Francisco, CA, USA*

WOLFGANG LÖSCHER • *Department of Pharmacology, Toxicology and Pharmacy, University of Veterinary Medicine, Hannover, Germany; Center for Systems Neuroscience, Hannover, Germany*

ANTHONY G. MARSON • *Institute of Translational Medicine, University of Liverpool, Liverpool, UK*

ALAN MORGAN • *Institute of Translational Medicine, University of Liverpool, Liverpool, UK*

NIKITA NIRWAN • *Neurobehavioral Pharmacology Laboratory, Department of Pharmacology, School of Pharmaceutical Education and Research (SPER), Jamia Hamdard, New Delhi, Delhi, India*

HEIDRUN POTSCHKA • *Institute of Pharmacology, Toxicology, and Pharmacy, Ludwig-Maximilians-University, Munich, Germany*

MOHD. FAROOQ SHAIKH • *Neuropharmacology Research Strength, Jeffrey Cheah School of Medicine & Health Sciences, Monash University Malaysia, Bandar Sunway, Selangor, Malaysia*

GRAEME J. SILLS • *Institute of Translational Medicine, University of Liverpool, Liverpool, UK; School of Life Sciences, University of Glasgow, Glasgow, UK*

KATARZYNA SOCAŁA • *Department of Animal Physiology and Pharmacology, Institute of Biological Sciences, Marie Curie-Sklodowska University, Lublin, Poland; Department of Animal Physiology and Pharmacology, Institute of Biological Sciences, Maria Curie-Skłodowska University, Lublin, Poland*

BRENNAN J. SULLIVAN • *Neuroscience Laboratory, Hugo Moser Research Institute at Kennedy Krieger, Baltimore, MD, USA*

MANJARI TRIPATHI • *Centre of Excellence for Epilepsy, A joint collaboration between All India Insitute of Medical Sciences (AIIMS) and National Brain Research Centre (NBRC), New Delhi, India; Department of Neurology, All India Institute of Medical Sciences, New Delhi, India*

RAJKUMAR TULSAWANI • *Defense Institute of Physiology and Allied Sciences, Defense Research and Development Organization, New Delhi, India*

DIVYA VOHORA • *Neurobehavioral Pharmacology Laboratory, Department of Pharmacology, School of Pharmaceutical Education and Research (SPER), Jamia Hamdard, New Delhi, India*

PREETI VYAS • *Neurobehavioral Phamracology Laboratory, Department of Pharmacology, School of Pharmaceutical Education and Research (SPER), Jamia Hamdard, New Delhi, India*

PIOTR WLAŹ • *Department of Animal Physiology and Pharmacology, Institute of Biological Sciences, Marie Curie-Sklodowska University, Lublin, Poland*

Part I

Models for Seizures and Epilepsy: An Overview

Chapter 1

Models of Seizures and Epilepsy: Important Tools in the Discovery and Evaluation of Novel Epilepsy Therapies

Wolfgang Löscher

Abstract

Models of seizures and epilepsy continue to play an important role in the early discovery of new therapies for the symptomatic treatment of epilepsy. Since the discovery of phenytoin in the 1930s, almost all antiseizure drugs (ASDs) have been identified by their effects in animal models, and millions of patients have profited from the successful translation of animal data into the clinic. However, several unmet clinical needs remain, including resistance to ASDs in about 30% of patients with epilepsy, adverse effects of ASDs that can reduce quality of life, and the lack of treatments that can prevent development of epilepsy in patients at risk following brain injury. This minireview, which serves as an introduction to the Neuromethods volume *Experimental and Translational Methods to Screen Drugs Effective Against Seizures and Epilepsy,*, briefly introduces the challenges and opportunities for in vivo and in vitro models of seizures and epilepsy and how such models may help to develop the next generation of drug therapies.

Key words Kindling, Status epilepticus, Antiseizure drugs, Anti-epileptic drugs, Epileptogenesis

1 Introduction

Epilepsy is one of the most common and disabling neurological disorders, characterized by unpredictable spontaneous recurrent seizures (SRS), which may affect all age groups [1]. Epilepsy or, more precisely, the epilepsies are a spectrum condition with a wide range of seizure types, ranging from nonconvulsive to convulsive, focal to generalized, and many more [2]. Also, the causes (or etiology) of the epilepsies vary widely from structural/metabolic (or acquired) to genetic to unknown [2]. In addition to SRS, patients with epilepsy may suffer from comorbidities, such as depression, anxiety, and cognitive impairment, and the social consequences of seizures and comorbidities [1]. In about 70% of patients, seizures can be controlled by symptomatic treatment with antiseizure drugs (ASDs; previously also termed anti-epileptic or anticonvulsant drugs), but one-third of patients will continue to

Divya Vohora (ed.), *Experimental and Translational Methods to Screen Drugs Effective Against Seizures and Epilepsy,* Neuromethods, vol. 167, https://doi.org/10.1007/978-1-0716-1254-5_1,
© Springer Science+Business Media, LLC, part of Springer Nature 2021

have uncontrolled seizures, which is referred to as drug resistance [1]. In patients with drug-resistant epilepsy, seizures can be fatal owing to direct effects on autonomic and arousal functions or owing to indirect effects such as drowning and other accidents [1]. Thus, there is an urgent need to develop more effective ASDs for those patients that are currently not controlled by treatment. Furthermore, despite the fact that cerebral insults, such as severe traumatic brain injury (TBI), brain tumors and stroke, or CNS infections, are associated with a substantially increased risk of epilepsy, nothing can be done to prevent epilepsy in patients at risk [3]. Thus, development of antiepileptogenic therapies that can be administered shortly after brain insults to prevent or modify the development of epilepsy and associated comorbidities is another unmet medical need [4].

For both, development of more effective antiseizure therapies and novel antiepileptogenic treatments, adequate in vitro and in vivo models of seizures and epilepsy are essential. Indeed, animal models of seizures and epilepsy have been important to advance our understanding of the fundamental neurobiology of epilepsy and to discover and develop more than 20 ASDs in the last some 30 years [5]. Compared to older ASDs, the pharmacokinetic and safety profiles of the newer drugs have improved measurably, but the problem of drug resistance has not yet been resolved. Similarly, there is still no approved treatment to prevent epilepsy in patients at risk, although some clinical studies with repurposed molecules, such as statins, vigabatrin, and levetiracetam, may suggest that such treatments exist [6].

The gaps in current epilepsy therapy have initiated rethinking of the strategies and preclinical methodologies used to identify novel medications [5]. As a visible result, the US National Institutes of Health (NIH)/National Institute of Neurological Disorders and Stroke (NINDS)-sponsored Anticonvulsant Screening Program (ASP), which since its initiation in 1975 spurred the development of many clinically used ASDs today, was critically assessed recently to become the Epilepsy Therapy Screening Program (ETSP), a name that better reflects the broad nature of emerging and recently validated treatment approaches for epilepsy [7]. In this new program, the use of simple seizure screening tests in normal (non-epileptic) rodents, such as the maximal electroshock seizure (MES) or 6-Hz tests, has been restricted to an initial identification (discovery) phase, while more laborious chronic models of epilepsy are used for subsequent drug evaluation in the identification and differentiation phases of the program [5, 7]. In this respect, it is important to consider that epilepsy is not a singular disease, but more accurately reflects heterogeneous diseases ("epilepsies"), so there is no ideal animal model that mirrors all the key features of even one human epileptic condition [5]. Thus, the core strength of the current staged and multipronged ETSP approach toward ASD

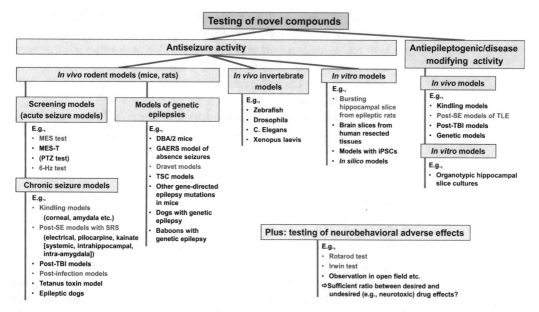

Fig. 1 Different categories of models of seizures and epilepsies and their use for either evaluation of antiseizure or antiepileptogenic/disease-modifying activities of treatments. Models used in the NIH/NINDS-sponsored ETSP are highlighted in red (see text). For more details see Löscher [8]. Abbreviations: *GAERS* genetic absence epilepsy rat from Strasbourg, *iPSC* induced pluripotent stem cell, *MES* maximal electroshock seizure, *MES-T* MES threshold, *PTZ* pentylenetetrazole, *SRS* spontaneous recurrent seizures, *TBI* traumatic brain injury, *TLE* temporal lobe epilepsy, *TSC* tuberous sclerosis complex

development lies in the use of multiple animal models, including the kindling model, and the inherent flexibility to capture promising drugs without "missing" potentially impactful agents [5].

In Fig. 1, several widely used in vivo and in vitro models of seizures and epilepsy are illustrated. Models used in the current version of the ETSP are highlighted. Several of these models can be used both in the search of novel ASDs *and* antiepileptogenic treatments. Many of the models shown in Fig. 1 are also used in the NIH/NINDS-sponsored ETSP [7]. Most of these models are described in detail in the present volume of Neuromethods; therefore, no details are discussed here. Of particular interest for drug discovery and differentiation are models of temporal lobe epilepsy (TLE), such as amygdala kindling or post-status epilepticus (post-SE) models, because most patients with this type of epilepsy do not sufficiently respond to current therapies and are thus used for first-in-man add-on therapies with novel ASDs [8]. Indeed, drug testing in amygdala-kindled rats correctly predicted the antiseizure efficacy of most novel compounds in patients with TLE [8].

One advantage of amygdala kindling and post-SE rodent models of TLE is that ASD-resistant and ASD-responsive individuals can be selected from large groups of kindled or epileptic animals [8, 9]. This allows both studying mechanisms of ASD resistance and developing more effective ASDs. According to PubMed, the amygdala-kindled rat was the first animal model of pharmacoresistant epilepsy described in the literature [10], subsequently followed by various other models of ASD-resistant epilepsy, including the 6-Hz model in mice and rats and chronic models with pharmacoresistant SRS (Fig. 1).

In addition to developing novel treatments for patients with TLE and other types of acquired epilepsies, new treatments for genetic epilepsies are important [11]. Over the past two decades, technological advances in genomics have led to the identification of nearly 1000 genes known to be associated with epilepsy [12, 13]. Indeed, epilepsy genes are revolutionizing our understanding and treatment of the disease [13]. By targeted mutagenesis in model systems, such as mice, zebrafish, flies, *C. elegans*, heterologous expression systems, human-induced pluripotent stem cells (iPSCs), and others, experimental test systems to search for novel, etiology-specific treatments ("precision medicine") for genetic epilepsies can be generated (Fig. 1) [11, 13]. Enthusiasm for precision medicine currently stems largely from discoveries from genetics about the causation of some of the rare, severe, typically early-onset epilepsies, including the developmental and epileptic encephalopathies. The list of genes carrying pathogenic rare variants is growing at a rapid pace. These discoveries have in some cases led to better understanding of disease biology, and, occasionally, rational treatment strategies have been devised, including better selection from existing ASDs or repurposing of drugs licensed but previously not for use in epilepsy, sometimes with dramatic responses [11, 14]. One relevant example is the recent approval of everolimus for treatment of refractory focal-onset seizures in patients with tuberous sclerosis complex (TSC). Everolimus is an immunosuppressive drug that acts by inhibiting the mammalian target of rapamycin (mTOR) [15]. In both mouse models of TSC and patients with this devastating disease, the mTOR signalling cascade is activated, which is thought to underlie seizures and other symptoms [15].

In conclusion, knowledge of the heterogeneity of epilepsy and epileptic seizures and how to simulate this heterogeneity by adequate models is a prerequisite when it comes to using such models for studying the neurobiology and pharmacology of the disease. The chapters in this volume of Neuromethods illustrate numerous relevant in vivo and in vitro models of seizures and epilepsy and describe laboratory protocols and step-by-step methods but also address methodological controversies and pitfalls in order to aid neuroscientists in experimentation. Thus, this volume may help scientists to develop future epilepsy therapies, whether they represent novel modifications aimed at traditional molecular targets or whether their origin is from a radically different understanding of the neurobiology of epilepsy.

References

1. Devinsky O, Vezzani A, O'Brien TJ, Jette N, Scheffer IE, De Curtis M, Perucca P (2018) Epilepsy. Nat Rev Dis Primers 4:18024
2. Berg AT, Berkovic SF, Brodie MJ, Buchhalter J, Cross JH, van Emde BW, Engel J, French J, Glauser TA, Mathern GW, Moshe SL, Nordli D, Plouin P, Scheffer IE (2010) Revised terminology and concepts for organization of seizures and epilepsies: report of the ILAE Commission on Classification and Terminology, 2005–2009. Epilepsia 51:676–685
3. Klein P, Dingledine R, Aronica E, Bernard C, Blümcke I, Boison D, Brodie MJ, Brooks-Kayal AR, Engel J Jr, Forcelli PA, Hirsch LJ, Kaminski RM, Klitgaard H, Kobow K, Lowenstein DH, Pearl PL, Pitkänen A, Puhakka N, Rogawski MA, Schmidt D, Sillanpää M, Sloviter RS, Steinhauser C, Vezzani A, Walker MC, Löscher W (2018) Commonalities in epileptogenic processes from different acute brain insults: do they translate? Epilepsia 59:37–66
4. Löscher W, Klitgaard H, Twyman RE, Schmidt D (2013) New avenues for antiepileptic drug discovery and development. Nat Rev Drug Discov 12:757–776
5. Rho JM, White HS (2018) Brief history of anti-seizure drug development. Epilepsia Open 3:114–119
6. Klein P, Friedman A, Hameed M, Kaminski R, Bar-Klein G, Klitgaard H, Koepp M, Jozwiak S, Prince D, Rotenberg A, Vezzani A, Wong M, Löscher W (2020) Repurposed molecules for antiepileptogenesis: missing an opportunity to prevent epilepsy? Epilepsia 61:359–386
7. Kehne JH (2017) National Institute of Neurological Disorders and Stroke (NINDS) Epilepsy Therapy Screening Program (ETSP). Neurochem Res 42:1894–1903
8. Löscher W (2016) Fit for purpose application of currently existing animal models in the discovery of novel epilepsy therapies. Epilepsy Res 126:157–184
9. Löscher W (2011) Critical review of current animal models of seizures and epilepsy used in the discovery and development of new antiepileptic drugs. Seizure 20:359–368
10. Löscher W, Jäckel R, Czuczwar SJ (1986) Is amygdala kindling in rats a model for drug-resistant partial epilepsy? Exp Neurol 93:211–226
11. Demarest ST, Brooks-Kayal A (2018) From molecules to medicines: the dawn of targeted therapies for genetic epilepsies. Nat Rev Neurol 14:735–745
12. Wang J, Lin ZJ, Liu L, Xu HQ, Shi YW, Yi YH, He N, Liao WP (2017) Epilepsy-associated genes. Seizure 44:11–20
13. Noebels J (2015) Pathway-driven discovery of epilepsy genes. Nat Neurosci 18:344–350
14. Moller RS, Hammer TB, Rubboli G, Lemke JR, Johannesen KM (2019) From next-generation sequencing to targeted treatment of non-acquired epilepsies. Expert Rev Mol Diagn 19:217–228
15. Wong M (2013) A critical review of MTOR inhibitors and epilepsy: from basic science to clinical trials. Expert Rev Neurother 13:657–669

Part II

Ex Vivo and In Vitro Models

Chapter 2

Protocol for Rodent Organotypic Hippocampal Slice Culture Model for Ex Vivo Monitoring of Epileptogenesis

Shabnam Ghiasvand, Jing Liu, Md. Joynal Abedin, and Yevgeny Berdichevsky

Abstract

Brain injury is one of the common causes of acquired epilepsy. Current treatments include anti-epileptic drugs that are anticonvulsant. However, around 30% of patients become resistant. Better understanding of epileptogenesis, the process in which the injury-induced cellular and subcellular modifications transform a healthy neuronal network to a malfunctioned state, has the potential to generate novel, more effective treatments. Creating a model that replicates various aspects of brain injury, and an experimental platform where the causality of these alterations can be investigated individually, has been a challenge. In the recent decade, organotypic hippocampal cultures were introduced as a posttraumatic epileptogenesis model that encompasses different aspects of the epileptogenesis process in a compressed time scale such that within 7 days, cultures become spontaneously epileptic. Here, in this chapter, we seek to elucidate and describe several methodologies that can be utilized to study epileptogenesis in vitro.

Key words Organotypic hippocampal cultures, Epilepsy, Epileptogenesis, Seizure, Traumatic brain injury, Electrophysiology, Optogenetics

1 Introduction

Explantation of slices of nervous tissue to an in vitro setting has been an attractive tool in the field of neuroscience for almost 30 years [1]. These slices can be dissected from different regions of the brain such as the hippocampus, cortex, and thalamus and cultured in conditions that replicate the body environment [2]. These slice cultures, when maintained at 37 °C with proper oxygenation, growth media, and on an appropriate substrate platform, can survive for several weeks, while preserving the cytoarchitecture of the origin of the tissue to a great degree. Therefore, they have become an important tool for many physiological investigations as well as for high-throughput pharmacological and electrophysiological experiments [3–5].

Divya Vohora (ed.), *Experimental and Translational Methods to Screen Drugs Effective Against Seizures and Epilepsy*, Neuromethods, vol. 167, https://doi.org/10.1007/978-1-0716-1254-5_2,

Since the hippocampus has a laminar structure with well-defined neuronal layers, it has been a favorable choice for many anatomical studies as the anatomical alterations are readily distinguishable and quantifiable. The hippocampus is subdivided into three main sub-regions including CA1, CA3, and the dentate gyrus (DG). Each sub-region contains unique neuronal layers and neuron types: the relatively small pyramidal neurons of CA1, large pyramidal neurons of CA3, and small and densely packed granule cells of DG [6]. Moreover, the hippocampus has been shown to be involved in various neurological diseases. Due to the critical role of the hippocampus for learning and memory, it is one of the structures affected early in Alzheimer's disease (AD) [7]. In Parkinson's disease, there is evidence to suggest that hippocampal degeneration occurs regardless of whether the disease is associated with dementia [8]. Other studies from patients with schizophrenia demonstrate that the hippocampus undergoes morphological and molecular changes that result in malfunctioning outputs [9]. Studies of patients with temporal lobe epilepsy (TLE) provide evidence of synaptic rewiring or sclerosis of the hippocampus [10–12]. The anatomical features and the involvement of the hippocampus in different diseases make the hippocampus a widely studied region of the brain and, in the case of in vitro studies, make organotypic hippocampal cultures (OHCs) a useful model for numerous experimental purposes.

Even though OHCs maintain their intrinsic cellular structure and organization to a large degree, different morphological studies provide evidence that due to the afferent cut introduced by the dissection, which results in disconnection from the entorhinal cortex, the surviving neurons sprout new connections [13]. The ability of hippocampal neurons to sprout new axons has been mostly studied in granule cells of mossy fibers [14–16]. Structural studies from pyramidal cells of CA1 and CA3 have also confirmed the ability of these neurons to sprout new axons and reorganize the hippocampal circuitry in vitro [17, 18]. The physiological consequences of axonal sprouting in these regions were investigated and showed an increase in the frequency of spontaneous excitatory postsynaptic potentials [19, 20].

It has been observed from local field potential recordings from OHCs that these cultures are characterized by the presence of spontaneous prolonged, high frequency spiking in the manner of seizures (spiking frequency of greater than 2 Hz, and lasting for at least 10 s), after approximately 7 days of culturing [21]. Based on the observed axonal sprouting and the developed hyperexcitability of the network, a hypothesis can be proposed that the connections made by these new projections result in an abnormal circuitry which leads to the pathological excitability of the network. In addition to loss of connections, these cultures undergo two distinct waves of neuronal cell death. The first wave is damage-driven and occurs during the first week of culturing as the result of the induced injury. The second wave is caused by the persistent ictal activity during the second and third weeks in vitro and can be abolished by activity inhibition [22]. Another feature observed in OHCs is the activation of microglial cells [23]. Cell death, apoptosis, and glial cell activation have been observed after traumatic brain injury (TBI) and in chronic epilepsy [24–26], and OHC model captures both phenomena [22, 23].

Brain injury is one of the leading causes of disability as well as one of the leading causes of epilepsy. Brain injury commonly results in neuronal loss and deafferentation between neuronal populations. Studies have shown there is a direct correlation between the severity of the injury and progression of the disorder [27–30].

Several obstacles exist in managing epilepsy clinically. The treatment of epilepsy is primarily focused on the use of anti-epileptic drugs (AEDs). AEDs are anticonvulsants and need to be taken continuously. However, approximately 30–40% of patients become resistant to currently available AEDs [31]. Additionally, patients who have undergone surgical treatment are only seizure-free 5 years afterward at a rate of $52 \pm 4\%$ [32]. The difficulty in treating epilepsy and the lifestyle constraints imposed by the unpredictability of seizures place great strains on patients and families, thereby significantly reducing the quality of life. The current lack of effective treatment is partly due to an incomplete understanding of the conditions necessary and lack of platforms for the study of epileptogenesis. Traumatic injury triggers pathological alterations in the brain such as cell death, inflammation, neurogenesis, axonal and dendritic reorganization, synaptogenesis, and changes in gene expression [33, 34]. It is hypothesized that these neurological modifications may push a healthy neuronal network closer to an epileptic state. Therefore, "epileptogenesis" is defined as a process by which a healthy network becomes epileptic. Epileptogenesis is then divided into three stages, first, the occurrence of injury; second, a latent phase when the transition from healthy network to the epileptic state takes place; and finally the state of pathological, recurrent seizures. The time from the incidence of the injury to development of the epilepsy varies from case to case and ranges from a few weeks to years. This latency period provides a potential therapeutic window where potential interventions might be given to prevent the progression of the disorder [35].

Among the various alterations that occur post-injury, aberrant axonal sprouting is the most ubiquitous feature that has been observed in many posttraumatic epilepsy models. Additionally, it has been reported from studies on the human patients with different epilepsy etiologies [34, 36, 37]. Axonal sprouting in response to injury has been thoroughly documented throughout the past century, beginning with Ramon Y Cajal reporting sprouting of neocortical neurons after injury and proposing it as a compensatory mechanism [18, 20, 38, 39].

Taken together, these alterations demonstrate the suitability and convenience of organotypic cultures as a model of posttraumatic epilepsy. OHCs develop epilepsy quickly and spontaneously without the need for pharmacological or electrical intervention. They provide a convenient platform in which chronic electrophysiological, optical, or molecular data may be collected. And, they accurately capture the important features of epileptogenesis [27, 40, 41].

Until about a decade ago, OHCs were not used as a model of posttraumatic epilepsy [20, 41]. Most studies have focused on induction-based models of seizures for in vitro study of epilepsy, where either chemical reagents, such as 4-aminopyridine (4-AP), or electrical kindling results in seizure generation [42].

In this chapter, we focus on different techniques that can be used to study epileptogenesis in OHCs. We begin by describing the process of generating and maintaining these cultures. Neuronal density and neuron number can be determined for neurotoxicity or neuroprotection studies by staining for NeuN, a marker for neuronal nuclei [43]. From the confocal stack taken from the entire body of the slices and via image processing techniques, the number of neurons, the density, and the morphology of neuronal layers may be obtained. In Fig. 1, one optical layer of different sub-regions of a hippocampal slice is shown.

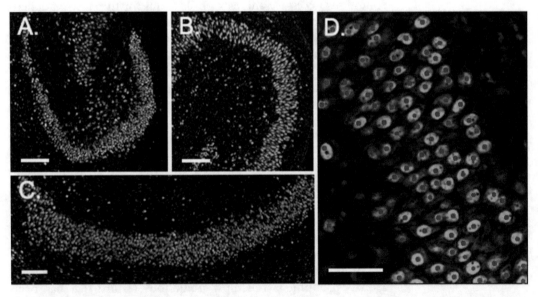

Fig. 1 Immunofluorescent staining of organotypic hippocampal cultures for NeuN, a neuronal marker which marks the nucleus of neurons. A single optical layer of confocal stacks from different sub-regions of the hippocampus is represented; (a). DG, (b). CA3, and (c). CA1. (d). Magnified view of a CA3 neuronal layer is shown. Neurons were detected by an automated cell counting code through ImageJ, and counted cell in one optical layer is shown in small red dots. Scale bars, 500 and 100 µm

Monitoring spontaneous seizure-like activity in OHCs is a powerful investigative tool that enables evaluation of the effect of different drugs and inhibitors. In order to examine the effects of the drugs on the activity of these cultures, two general screening techniques are described: (1) local field potential (LFP) recordings via a tungsten microwire placed underneath the culture and (2) optical recordings through expression of jRGECO1a, a genetically encoded calcium indicator which reflects activity-dependent changes in intracellular $[Ca^{2+}]$. Both techniques enable assessment of the frequency and duration of seizures.

Chronic electrical recordings of OHCs are a necessary tool for the evaluation of AEDs and the monitoring of epileptogenesis. Microwire recordings have been used for decades in in vivo settings [44–46] and neural implants and are easily integrated into culture dishes for chronic electrophysiological data collection in vitro. Figure 2 represents the experimental layout for chronic screening of electrical activities of OHCs in response to different drug applications. We have used this technique to investigate culture responses to a library of kinase inhibitors by measuring the seizure frequency, duration, and electrographic load [4].

Spontaneous activity of cultures can also be recorded optically through expression of jRGECO1a (red fluorescent genetically encoded $[Ca^{2+}]$ indicator) via AAV vectors to the culture media. It takes the cultures 6–8 days after AAV application to express the calcium indicator [47, 48]. jRGECO1a consists of mApple, a red fluorescent protein (RFP), fused to calmodulin and the M13 peptide. In the presence of light, during action potentials when the intracellular calcium concentration rises, calmodulin undergoes a conformational change to associate with M13 which modifies the chromophore environment of mApple and increases RFP brightness [49]. Optical recordings present a tradeoff relative to electrical recordings. Optical recordings offer superior spatial resolution, even relative to recordings performed on multi-electrode arrays, by enabling continuous observation of the entire culture. However, due to relatively slow kinetics of post-activity decrease of intracellular $[Ca^{2+}]$ and of jRGECO1a fluorescence, temporal resolution is reduced compared to electrical recordings.

We also describe molecular assays such as lactate and lactate dehydrogenase (LDH) as markers of seizure-like activity and cell death for high-throughput evaluation of potential therapeutics. Lactate and lactate dehydrogenase (LDH) are two biomarkers that can be used to assess the effectiveness of anticonvulsants on both seizures and seizure-induced cell death. Lactate dehydrogenase is released into culture growth medium when the cell membrane is compromised and can therefore be used as a marker for neuronal cell death [50]. In addition, lactate concentration has been shown to increase in the culture medium following the occurrence of seizure-like activity [33]. Moreover, it has been shown in

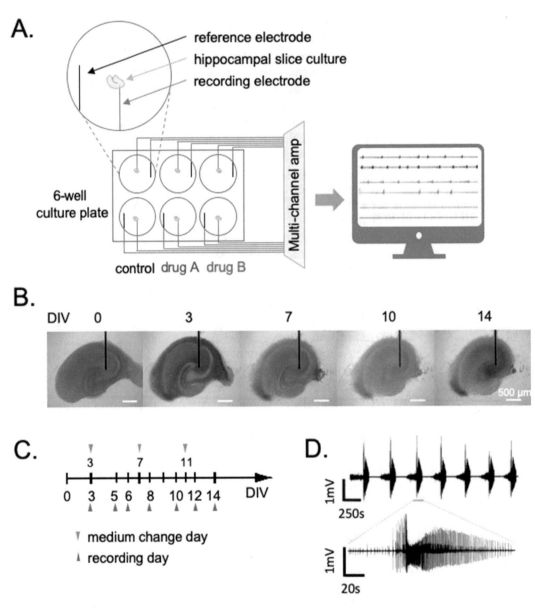

Fig. 2 Microwire-based chronic recording system. (**a**) Microwire-integrated culture plate and the data acquisition system. The recording electrode is centered in the well where the slice culture will be positioned, and the reference electrode (with insulation layer removed at the tip) is placed on the side. (**b**) Brightfield images from 0 to 14 DIV show that microwire electrodes do not compromise the survival of organotypic cultures. The thin black line in each image is the recording electrode. Scale bars, 500 μm. (**c**) Long-term recording schedule for a 2 week culture experiment. (**d**) Sample electrical recording taken with a microwire recording electrode. Bottom trace is a zoomed-in view of one seizure, detected by MATLAB according to the set criteria for seizure detection described in the text

patients with complex partial seizures, who were resistant to AEDs, there is a direct relationship between extracellular lactate content and ictal activity at the site of the seizure [51].

2   Materials

2.1   Materials Required for Explant of Organotypic Hippocampal Cultures

1. 35 mm tissue culture petri-dishes or 6-well plates (Falcon) can be used as the substrate for the OHCs, depending on the experimental setup requirements.

2. Poly-D-lysine (PDL, Sigma Aldrich) is applied to provide proper attachment.

3. Dissection solution contains Gey's balanced salt solution with 0.3 mM kynurenic acid (KYNA) and 6 mg/mL glucose.

4. McIlwain tissue chopper (Mickle Laboratory Eng. Co., Surrey, United Kingdom) is used to divide the hippocampus into slices at desired thickness.

5. Culture medium is a serum-free solution that consists of Neurobasal-A/B27, 30 μg/mL gentamicin, and 0.5 mM GlutaMAX (Invitrogen).

2.2   Immuno-fluorescent Staining

1. 4% paraformaldehyde in phosphate-buffered saline (PBS) is used as fixation solution.

2. Permeabilization is done with 0.3% Triton X-100 (Sigma-Aldrich) in phosphate-buffered saline (PBS).

3. Blocking buffer consists of 10% goat serum in 0.05% Triton X-100 in PBS (PBST).

4. Anti-NeuN conjugated to Alexa Fluor 555 (Millipore) is used as the neuronal nuclei marker.

5. Glass coverslips are used to allow confocal imaging with Zeiss confocal microscope with $40\times$ or $25\times$ objectives (Zeiss LSM 510 META, Germany).

6. Fluorogel is used to mount the cultures on coverslips (*see* **Note 1**).

2.3   Microwire Culture Plates for LFP Recordings

1. PFA-coated tungsten wires (bare diameter $= 50.8$ μm, coated diameter $= 101.6$ μm, A-M systems Inc.) are utilized as the recording and reference electrodes.

2. 6-well culture plates are used as the substrates for culture explantation.

3. All the electrode placement achieved by using an implant grade silicone adhesive (4300 RTV, Bluestar Silicones).

2.4   AAV Infection for Genetically Encoded Calcium Indicator

1. 35 mm tissue culture petri-dishes are used as OHC substrate for the purpose of optical recordings.

2. pAAV.Syn.NES-jRGECO1a.WPRE.SV40 is a gift from Douglas Kim & GENIE Project (Addgene plasmid #100854; http://n2t.net/addgene:100854; RRID: Addgene_100854) and is applied to culture media for optical recording purposes.

2.5 Lactate and LDH Assays

1. L-Lactate Assay Kit I (Eton Bioscience, SKU#1200014002) is used for lactate assay.

2. Cytotoxicity Detection Kit (Roche Diagnostics, CAT#11644793001) is used for LDH assay.

3 Methods

3.1 Explant of Organotypic Hippocampal Cultures

1. 6-well culture plates or 35 mm petri-dishes are coated with PDL 2 days prior to dissection day and kept in a humidified 37 °C incubator with 5% CO_2.

2. The PDL is then removed and the plates are washed with sterile water $(3\times)$; at the second wash sterile water is left in the wells for at least 20 min.

3. The wells are then filled with Neurobasal A/B27 medium and incubated overnight or for at least 3 h before dissection.

4. Sterile water is added to the spacing between wells. 35 mm petri-dishes are placed within a larger dish alongside a single extra 35 mm petri-dish filled with sterile water and kept with tissue culture dishes only to provide proper humidity.

5. At the dissection day, 0 day in vitro (0 DIV), the substrate is then washed 2 times with Gey's solution and again transferred back to the incubator. From this step forward the plate is kept over an empty 6-well plate avoiding direct contact with the incubator surface to avoid uneven heating.

6. Hippocampi of Sprague-Dawley rats of post-natal days 7–8 are transferred to ice-cold dissection solution.

7. Slices with 350 μm thickness are made using tissue chopper.

8. Slices are then placed onto PDL coated 6-well tissue culture plates or 35 mm tissue culture petri-dishes.

9. 625 μL of culture media is added to each well, and dishes are transferred to the incubator.

10. After 1 day in incubation dishes are transferred onto a rocking platform.

11. Culture media are exchanged twice per week with fresh warm media.

3.2 Immuno-fluorescent Staining

1. At the desired DIV, cultures are washed briefly with room-temperature PBS $2\times$.

2. Cultures are then fixed with 4% paraformaldehyde for 1 h at RT.

3. After fixation, using a scalpel cultures are scraped off of the substrate and transferred to a 48-well plate.

4. Cultures are again washed shortly with PBS 2×; 200 μL per well is sufficient to cover slices.

5. Permeabilization is done through exposure to 0.3% Triton X-100 in PBS for 2 h on a shaking platform at RT.

6. Cultures are then blocked in 10% goat serum in PBST for 1 h.

7. Anti-NeuN at 1:500 in blocking buffer is then added to the wells and incubated at 4 °C on a rotating platform for 48 h.

8. Cultures are washed 4×, each time 20 min on a shaking platform at RT with 1% goat serum in PBST.

9. Last wash includes 3×, each time 15 min on a shaking platform at RT with PBS to wash away all the remaining serum.

10. Lastly, cultures are transferred to coverslips, and residual solution is extracted carefully by a Kimwipe while avoiding contact with the slice.

11. One drop of Fluorogel is added on the slice, and another coverslip is used to mount the slice in between the coverslips. With this technique fixed cultures can be kept in the refrigerator for up to 3 weeks.

12. Optical stacks were imaged from the entire thickness of the cultures using Zeiss confocal microscope with 40× or 25× objectives (Zeiss LSM 510 META, Germany). Optical slices were taken with 1 μm intervals.

13. Optical stacks can then be analyzed with ImageJ in order to assess the health of the cultures. Neuronal counting is done with slightly modified version of the existing 3D watershed technique from [52] to properly detect the nuclei of pyramidal cells in CA1 and CA3 layers and granule cells in DG neuronal layer (*see* **Note 1**).

3.3 Electro-physiological Recordings and Analysis

1. On the day of dissection, slices made by chopper are centered on the recording electrode such that the tip is underneath CA3 or CA1 neuronal layers.

2. Maintaining the slices in incubator is no different from the abovementioned protocol in Subheading 3.1.

3. On the day of recording, 6-well tissue culture plates are transferred to a recording chamber (Bioscience Tools) connected to a temperature controller maintaining temperature at 37 °C and a blood gas providing 5% CO_2, 21% oxygen, and balanced nitrogen (Airgas) for LFP recordings.

4. Extracellular field potentials are recorded for 45 min via high-impedance multiple-channel pre-amplifier stage (PZ2–64, Tucker Davis Technologies) connected to a RZ2 amplifier (Tucker Davis Technologies).

5. The recorded signal is analyzed with MATLAB. The first 15 min of each recording is not included in the analysis, in order to give cultures a period of time to adjust to the conditions in the recording chamber.

6. Signals are filtered with a band-pass filter (1 Hz to 3 kHz, gain × 1000) and sampled at 6 kHz.

7. A threshold of 65 μV is applied to identify population activity. Electrographic seizures are identified as paroxysmal events with event frequency greater than 2 Hz for a period of at least 10 s.

8. Recorded signals are analyzed using OpenX software (Tucker Davis Technologies) and MATLAB (MathWorks).

3.4 Assessment of Drug Application Using LFP Recordings

1. LFP recordings are collected for 45 min (in a similar process as noted before).

2. Drug (such as 50 μM of phenytoin) is then applied in culture media.

3. After 15 min of incubation, the culture is transferred to the recording chamber, and electrical signals are recorded for additional 45 min (culture is kept in the media containing the drug).

4. Lastly, culture media are washed out and replaced with fresh medium without the drug.

5. After a 15 min incubation period on the rocker platform, LFP recordings are collected again (denoted as washout trace in Fig. 3a).

3.5 Optical Recordings and Image Processing

1. jRGECO1a is applied to the culture medium on 0 DIV after 1 h incubation at a final concentration of about 10^{10} genome copies/mL.

2. Half of the medium is replaced with fresh culture media on DIV 3, and subsequently the whole culture medium is changed twice per week.

3. It takes the cultures 6–8 days after AAV application to express the calcium indicator [47, 48]; therefore, optical recordings can be collected after about a week in vitro.

4. 35 mm petri-dishes are placed into the mini-incubator where they are maintained at 37 °C and supplemented with blood gas, over a fluorescent inverted microscope stage (Olympus).

5. A CCD camera and a 4× objective are used to record fluorescent changes at frame rates ranging from 1.59 to 0.16 s/frame (~0.6–6 Hz). Activity is recorded for about 15–30 min long. Similar to the electrical recordings, since the cultures are no longer placed on a rocking stage, the recordings are limited to less than 1 h.

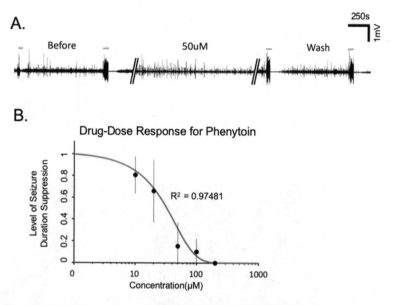

Fig. 3 Electrical recordings from OHCs in a short-term drug application protocol. (a) Sample electrical recordings before phenytoin application, after applying 50 μM drug, and after washout are shown. Between each recording, the cultures were placed back into the incubator for 15 min. (b) Total duration of seizures before, during, and after the drug application was calculated. Effectiveness of drug was measured by the level of seizure duration suppression which was defined as the ratio of total duration of seizure when the drug is applied, to the average of total seizure duration before drug application and after-wash recording

6. Initially, using ImageJ "ROI Manager," different regions of interest are selected for different cultures, and mean gray values are measured for each frame and recorded as a .txt file.

7. Values are then imported via MATLAB, and the baseline is calculated for the optical signal using the asymmetric least square smoothing method [53].

8. Signal to baseline ratio is calculated by the following equation:

$$\Delta F/F = \frac{(F(t) - F0)}{F0}, F0 \text{ is baseline} \tag{1}$$

9. In order to extract the paroxysmal activities, a simple thresholding method is used. The threshold is greater than 5% $\Delta F/F$. For seizure-like event detection, we use the same criteria as we use for electrical recordings; the optical signal above the threshold must last for at least 10 s to be counted as a seizure.

10. In order to make $\Delta F/F$ videos, 20–100 frames of each recording in which the culture is in a quiet state and inactive (lowest

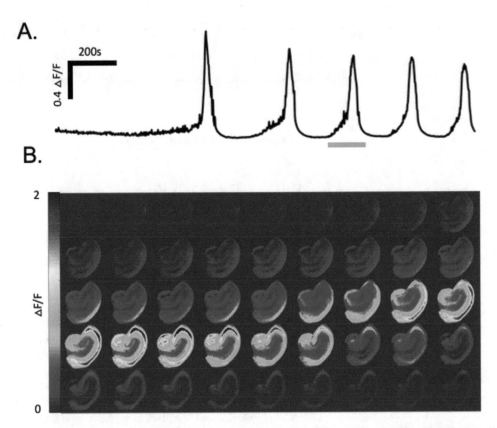

Fig. 4 Optical measurement, via jRGECO1a, of spontaneous seizure-like activity of OHCs. (**a**) Sample optical trace. (**b**) Spatial extent of paroxysmal activity in a culture before, during, and after a seizure is provided as a sequence of frames. Intervals between frames are approximately 3.8 s. The light-blue bar in (**a**) represents this time window

brightness) are selected and averaged into a baseline frame. The baseline frame is then subtracted from each frame pixel by pixel to calculate ΔF; the difference is divided by baseline for each pixel [54] (Fig. 4).

3.6 Chronic Measurement of Ictal Activity and Cell Death Through Molecular Assays

3.6.1 Lactate Assay Protocol

1. 100 μL of culture media is collected from slices on 3, 7, 10, 14, and 17 DIVs and stored in −80 °C.

2. 50 μL L-Lactate standards (0, 0.05, 0.1, 0.2, 0.4, 0.8, 1.6, 3.2 mM) are added to each well of a 96 well-plate.

3. 10 μL of previously collected culture media of each DIV is added to separate wells.

4. Then 40 μL dH$_2$O is added to dilute the sampled culture media.

5. 50 μL of L-Lactate assay solution is added to all the wells containing L-Lactate standards and test samples.

6. Plate is covered with aluminum foil and is incubated for 30 min at 37 °C.

7. The absorbance level is measured at 490 nm using a microplate reader.

3.6.2 Lactate
Dehydrogenase (LDH)
Assay Protocol

1. 50 μL of fresh culture medium is added to three wells of a 96-well plate as blank samples.

2. 50 μL of collected culture supernatants is added to separate wells.

3. 50 μL working solution (1:45 Catalyst: Dye solution, available with assay kit) is added to all the wells containing blank and test samples.

4. Plate is then fully covered with aluminum foil and incubated for 1.5 h at 37 °C.

5. The absorbance level is measured at 490 nm using a microplate reader.

4   Discussion

We have used OHCs as a PTE model and conducted different experimental procedures in order to investigate epileptogenesis in vitro. Health of the OHCs has been assessed by immunostaining for anti-NeuN, and we estimated the number of neurons in a relatively healthy culture to be around 30,000 neurons. The different sizes of the pyramidal neurons at different sub-regions of the hippocampus can also be visualized.

These cultures are capable of producing spontaneous seizures of variable shapes and durations at different frequencies from cultures to cultures and from DIV to DIV. We looked at these paroxysmal behaviors through two recording setups: (1) microwire recordings that enabled us to collect local field potentials from hippocampal slices with high temporal resolution and (2) calcium recordings through jRGECO1a expression via AAV infection which allowed us to capture the paroxysmal activities optically with high spatial resolution.

LFP recordings can be taken on different days in vitro, and the recording duration is recommended to be limited to 1 h, since the cultures are no longer on the rocking platform. Since epileptogenesis is a progressive process, chronic recordings provide the opportunity to explore evolution of spontaneous activity. Alternatively, effects of different drugs at different concentrations can be evaluated on the same culture by acute exposures to desired concentrations of drugs. Therefore, depending on the purpose of the research, two approaches can be taken using the same technique of microwire LFP recordings: chronic drug application or short-term exposure to a desired drug. Sample electrical recordings from a culture treated with short-term exposure to phenytoin, a widely prescribed AED that causes use-dependent inhibition of voltage gated sodium channels on neuronal cell membranes [55], are represented in Fig. 3a. From LFP recordings, epileptogenesis can be investigated through considering different factors: (1) number of seizures occurring within a specified time window; (2) length of each seizure, and/or total time culture is seizing during the recording time window; and (3) the inter-ictal activities. On average, these OHCs experience their first spontaneous seizure by 7 DIV.

Even though the LFP signals carry high temporal resolution information about electrical activities of these neuronal networks, they lack spatial information. Therefore, another recording approach was taken to compensate for this factor. With optical recordings through tracing calcium activity of these cultures, the seizure onset zone and propagation properties can be studied. For instance, in the seizure depicted in Fig. 4b, the seizure appears to initiate in the subiculum/CA1 sub-regions and propagates to the rest of the slice and displays the greatest intensity in CA3.

Two molecular assays were introduced as two biomarkers for cell death and ictal activity (LDH and lactate, respectively). With these two assays, chronical screening of epileptogenic progression from the cultures can also be monitored. LDH reaches a peak value due to cell death that occurs in response to the dissection; therefore, the supernatants collection can be initiated after the first media exchange (after the LDH due to dissection damage has been washed out). Another peak in the LDH level is found on later DIVs which is representative of the second wave of cell death that occurs as a result of ictal activity. Berdichevsky et al. demonstrated that cultures with suppressed seizures showed lower LDH release than seizing cultures. Since the amount of lactate in culture supernatants is proportional to seizure-like activity, drug candidates that successfully suppressed seizures could be identified [56].

5 Notes

The electrical behavior of OHCs, such as the frequency and duration of ictal activities, can be quite variable even for an individual culture, from DIV to DIV, or even within a 1 h recording. However, both optical and electrical techniques described here enable multiple measurements from the same culture on different DIVs, thus decreasing the number of required cultures.

1. Experiments must be interpreted in the context of the health of the cultures. One method, discussed earlier, for evaluating the health of cultures is staining for anti-NeuN and fluorescent imaging. However, this method is only available at the end of the experiment and after fixation. In some experimental setups, the quality of cultures is required to be known during the period of culturing and before fixation, in order to be able to interpret the collected data from OHCs and decide the subsequent procedures in experiment. There are

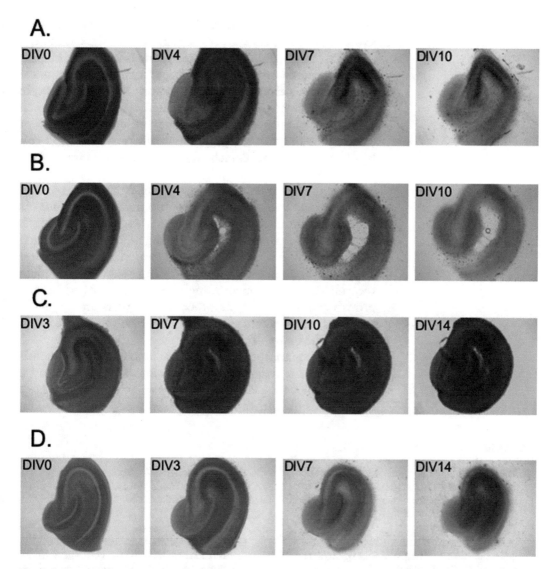

Fig. 5 Culture health assessment from early day in vitro brightfield images. (**a**) Brightfield images taken with 4× objective at different day in vitro from a culture that experienced excessive amount of distress at CA3 sub-region. The impact is apparent from DIV4; it is both deformed and the tissue is darker locally. (**b**) Representative brightfield images taken from a culture that experienced excessive shear that caused the tissue to break apart at the connective area between sub-regions. (**c**) Example of a culture exposed to toxic concentration of drug; first application of drug was done on DIV3 and the toxicity damaged the culture. Visual features here are tissue deformation, dark color, perforation, and loss of proper attachment. (**d**) Images of a healthy culture throughout 14 days of in vitro culturing

morphological features that can be easily observed with a standard inverted microscope with either brightfield or phase contrast microscopy. From brightfield images taken with 4× objective and solely from the morphological observation, the health quality of the culture can be surmised (Fig. 5). If the culture is abnormally dark or has spotted dark regions, it may be that during the dissection the culture experienced an excessive amount of pressure or damage at the darker regions (Fig. 5a). In general, dark-colored tissue is indicative of dead cells. In drug discovery experiments, at toxic concentrations it is quite common to obtain cultures with dark colors, and in some cases the culture disintegrates and collapses (Fig. 5c). If the culture shrinks over time, it may be due to insufficient adhesion to the substrate, and eventually the culture will collapse onto itself and detach. Another morphological feature that can help with evaluating the culture quality is the axon sprouting which can be observed through phase contrast images taken from the substrate surrounding the culture. As discussed in the introduction section of this chapter, the OHCs are capable of sprouting axons following injury. This sprouting starts as early as 1 day in vitro and can continue for almost 2 weeks. The presences of long and dense axons surrounding the culture are indicative of high culture quality and healthy growth conditions. On the other hand, if there are few axons surrounding a culture, it is likely that only a few neurons survived the injury of dissection.

Even though obtaining high-quality cultures can be challenging at the beginning of working with OHCs, after 1–3 months of experience one should achieve high-quality cultures reliably with around 90% consistency.

References

1. Gähwiler BH, Capogna M, Debanne D, McKinney RA, Thompson SM (1997) Organotypic slice cultures: a technique has come of age. Trends Neurosci 20(10):471–477

2. Gähwiler BH (1981) Organotypic monolayer cultures of nervous tissue. J Neurosci Methods 4(4):329–342. https://doi.org/10.1016/0165-0270(81)90003-0

3. Liu J, Sternberg AR, Ghiasvand S, Berdichevsky Y (2019) Epilepsy-on-a-chip system for antiepileptic drug discovery. IEEE Trans Biomed Eng 66(5):1231–1241. https://doi.org/10.1109/TBME.2018.2871415

4. Liu J, Schenker M, Ghiasvand S, Berdichevsky Y (2019) Kinase inhibitors with antiepileptic properties identified with a novel in vitro screening platform. Int J Mol Sci 20(10):2502

5. Cho S, Wood A, Bowlby MR (2007) Brain slices as models for neurodegenerative disease and screening platforms to identify novel therapeutics. Curr Neuropharmacol 5:19–33

6. Knowles WD (1992) Normal anatomy and neurophysiology of the hippocampal formation. J Clin Neurophysiol 9(2):252–263

7. Price DA, Scheff SW (2003) Synaptic pathology in Alzheimer's disease: a review of ultrastructural studies. Neurobiol Aging 24(8):1029–1046

8. Laakso MP, Partanen K, Riekkinen P et al (1996) Hippocampal volumes in Alzheimer's disease, Parkinson's disease with and without dementia, and in vascular dementia an MRI study. Neurology 46(3):678–681

9. Harrison PJ (2004) The hippocampus in schizophrenia: a review of the neuropathological evidence and tis pathophysiological implications. Psychopharmacology 174(1):151–162

10. Berkovic SF, Andermann F, Olivier A et al (1991) Hippocampal sclerosis in temporal lobe epilepsy demonstrated by magnetic resonance imaging. Ann Neurol 29(2):175–182

11. Dinkelacker V, Valabregue R, Thivard L et al (2015) Hippocampal-thalamic wiring in medial temporal lobe epilepsy: enhanced connectivity per hippocampal voxel. Epilepsia 56(8):1217–1226

12. Blumcke I, Thom M, Wiestler OD (2002) Ammon's horn sclerosis: a maldevelopmental

disorder associated with temporal lobe epilepsy. Brain Pathol 12(2):199–211

13. Gutierrez R, Heinemann U (1999) Synaptic reorganization in explanted cultures of rat hippocampus. Brain Res 815:304–316

14. Routbort MJ, Bausch SB, McNamara JO (1999) Seizures, cell death, and mossy fiber sprouting in kainic acid-treated organotypic hippocampal cultures. Neuroscience 94 (3):755–765

15. Coltman B, Eakley E, Shahak A et al (1995) Factors influencing mossy fiber collateral sprouting in organotypic slice cultures of neonatal mouse hippocampus. J Comp Neurol 362:209–222

16. Laurberg S, Zimmer J (1981) Lesion-induced sprouting of hippocampal mossy fiber collaterals to the fascia dentata in developing and adult rats. J Comp Neurol 200:433–459

17. Anderson K, Scheff S, Dekosky S (1986) Reactive synaptogenesis in hippocampal area CA1 of aged and young adult rats. J Comp Neurol 252:374–384

18. Takahashi DK, Gu F, Parada I, Vyas S, Prince DA (2016) Aberrant excitatory rewiring of layer V pyramidal neurons early after neocortical trauma. Neurobiol Dis 91:166–181

19. Buckmaster PS, Dudek FE (1997) Network properties of the dentate gyrus in epileptic rats with hilar neuron loss and granule cell axon reorganization. Am Phys Soc 77:2685–2696

20. Dyhrfjeld-Johnsen J, Berdichevsky Y, Swiercz W, Sabolek H, Staley KJ (2010) Interictal spikes precede ictal discharges in an organotypic hippocampal slice culture model of epileptogenesis. J Clin Neurophysiol 27 (6):418–424. https://doi.org/10.1097/WNP.0b013e3181fe0709

21. McBain CJ, Boden P, Hill RG (1989) Rat hippocampal slices "in vitro" display spontaneous epileptiform activity following long-term organotypic culture. J Neurosci Methods 27:35–49

22. Berdichevsky Y, Dzhala V, Mail M, Staley KJ (2012) Neurobiology of disease interictal spikes, seizures and ictal cell death are not necessary for post-traumatic epileptogenesis in vitro. Neurobiol Dis 45(2):774–785

23. Stence N, Waite M, Dailey ME (2001) Dynamics of microglial activation: a confocal time-lapse analysis in hippocampal slices. Glia 33:256–266

24. Hernandez-Ontiveros DG, Tajiri N, Acosta S et al (2013) Microglia activation as a biomarker for traumatic brain injury. Front Neurol 4:30

25. Henshall DC (2007) Apoptosis signaling pathways in seizure-induced neuronal death and epilepsy. Biochem Soc Trans 35(2):421–423

26. Rink A, Fung KM, Trojanowski JQ et al (1995) Evidence of apoptotic cell death after experimental traumatic brain injury in the rat. Am J Pathol 147(6):1575–1583

27. Pitkänen A, Immonen R (2014) Epilepsy related to traumatic brain injury. Neurotherapeutics 11(2):286–296

28. Lowenstein DH (2009) Epilepsy after head injury: an overview. Epilepsia 50(SUPPL. 2):4–9

29. Christensen J (2012) Traumatic brain injury: risks of epilepsy and implications for medicolegal assessment. Epilepsia 53(SUPPL. 4):43–47

30. Annengers JF, Rocca AW (1998) A population-based study of seizures after traumatic brain injuries. N Engl J Med 338 (20-4):8–12

31. Temkin NR (2001) Antiepileptogenesis and seizure prevention trials with antiepileptic drugs: meta analysis of controlled trials. Epilepsia 42(4):515–524

32. De Tisi J, Bell GS, Peacock JL et al (2011) The long-term outcome of adult epilepsy surgery, patterns of seizure remission, and relapse: a cohort study. Lancet 378:1388–1395

33. Berdichevsky Y, Dryer AM, Saponjian Y et al (2013) PI3K-Akt signaling activates mTOR mediated epileptogenesis in organotypic hippocampal culture model of post-traumatic epilepsy. J Neurosci 33(21):9056–9067

34. Pitkänen A, Lukasiuk K (2009) Molecular and cellular basis of epileptogenesis in symptomatic epilepsy. Epilepsy Behav 14(1 SUPPL. 1):16–25

35. Goldberg EM, Coulter DA (2014) Mechanism of Epileptogenesis: a convergence on neural circuit dysfunction. Nat Rev Neurosci 14 (5):337–349

36. Salin P, Tseng GF, Hoffman S, Parada I, Prince DA (1995) Axonal sprouting in layer V pyramidal neurons of chronically injured cerebral cortex. J Neurosci 15(12):8234–8245

37. Kharatishvili I, Nissinen JP, McIntosh TK, Pitkänen A (2006) A model of posttraumatic epilepsy induced by lateral fluid-percussion brain injury in rats. Neuroscience 140 (2):685–697

38. Sutula T, Cascino G, Cavazos J, Parada I, Ramirez L (1989) Mossy fiber synaptic reorganization in the epileptic human temporal lobe. Ann Neurol 26:321–330

39. Mckinney RA, Debanne D, Gähwiler BH et al (1997) Lesion-induced axonal sprouting and hyperexcitability in the hippocampus in vitro: implications for the genesis of posttraumatic epilepsy. Nat Med 3:990–996

40. Berdichevsky Y, Glykys J, Dzhala V et al (2017) Organotypic hippocampal slice cultures as a

model of posttraumatic epileptogenesis. In: Models of seizures and epilepsy; chapter 21. Elsevier, London, pp 301–311

41. Berdichevsky Y, Sabolek H, Levine JB, Staley KJ, Yarmush ML (2009) Microfluidics and multielectrode array-compatible organotypic slice culture method. J Neurosci Methods 178:59–64. https://doi.org/10.1016/j.jneumeth.2008.11.016

42. Kandratavicius L, Lopes-aguiar C, Ruggiero RN, Umeoka EH (2014) Animal models of epilepsy: use and limitations. Neuropsychiatr Dis Treat 10:1693–1705

43. Wolf HK, Buslei R, Schmdt-Kastner R et al (1996) NeuN: a useful neuronal marker for diagnostic histopathology. J Histochem Cytochem 44(10):1167–1171

44. Williams JC, Rennaker RL, Kipke DR (1999) Long-term neural recording characteristics of wire microelectrode arrays implanted in cerebral cortex. Brain Res Protocol 4:303–313

45. Kralik JD, Dimitrov DF, Krupa DR et al (2001) Techniques for long-term multisite neuronal ensemble recordings in behaving animals. Methods 25:121–150

46. Sankar V, Patrick E, Dieme R et al (2014) Electrode impedance analysis of chronic tungsten microwire neural implants: understanding abiotic vs biotic contribution. Front Neuroeng 7: Article 13. https://doi.org/10.3389/fneng.2014.00013

47. Lillis XKP, Wang Z, Mail M et al (2015) Evolution of network synchronization during early Epileptogenesis parallels synaptic circuit alterations. J Neurosci 35(27):9920–9934. https://doi.org/10.1523/JNEUROSCI.4007-14.2015

48. Keir SD, House SB, Li J, Xiao X, Gainer H (1999) Gene transfer into hypothalamic organotypic cultures using an adeno-associated virus vector. Exp Neurol 316:313–316

49. Dana H, Mohar B, Sun Y et al (2016) Sensitive red protein calcium indicators for imaging neural activity. elife 5:e12727

50. Lobner D (2000) Comparison of the LDH and MTT assays for quantifying cell death: validity for neuronal apoptosis? J Neurosci Methods 96 (2):147–152

51. During MJ, Fried I, Leone P, Katz A, Spencer DD (1994) Direct measurement of extracellular lactate in the human hippocampus during spontaneous seizures. J Neurochem 62 (6):2356–2361

52. Bindokas V, 17-Sep-2014. [Online]. Available: https://digital.bsd.uchicago.edu/%5Cimagej_macros.html

53. Baek S-J, Park A, Ahn Y-J, Choo J (2015) Baseline correction using asymmetrically reweighted penalized least squares smoothing. Analyst 140(1):250–257. https://doi.org/10.1039/c4an01061b

54. Namiki S, Norimoto H, Kobayashi C, Nakatani K, Matsuki N, Ikegaya Y (2013) Layer III neurons control synchronized waves in the immature cerebral cortex. J Neurosci 33 (3):987–1001. https://doi.org/10.1523/JNEUROSCI.2522-12.2013

55. Temkin NR, Dikmen SS, Wilensky AJ et al (1980) A randomized, double-blind study of phenytoin for the prevention of post-traumatic seizures. N Engl J Med 323:497–502

56. Berdichevsky Y, Saponjian Y, Park K et al (2016) Staged anticonvulsant screening for chronic epilepsy. Ann Clin Transl Neurol 3 (12):908–923. https://doi.org/10.1002/acn3.364

57. Ramirez JJ (2001) The role of axonal sprouting in functional reorganization after CNS injury: lessons from the hippocampal formation. Restor Neurol Neurosci 19(3–4):237–262. http://www.ncbi.nlm.nih.gov/pubmed/12082224

Chapter 3

Methods for the Screening of New Chemical Entities for Deciphering Neuroinflammatory and Associated Pathways in Seizures: An In Vitro Perspective

Preeti Vyas, Rajkumar Tulsawani, and Divya Vohora

Abstract

In the past few years, several conventional neuroinflammation-mediated epilepsy models have been modified to correlate the same with the actual clinical pathophysiological pattern. These models are being used for the screening of new antiseizure drugs as well as for the evaluation of associated cell-signaling events. Considering these applications, in this chapter, we describe a novel method for the screening of antiseizure molecules using an in vitro model based on neuronal inflammation in mouse hippocampal (HT22) cells. Based on our previous in vivo work, we propose the use of lipopolysaccharide as an inflammatory trigger with conventional pilocarpine model of seizures. Combination of these two can also help in the elucidation of the cell-signaling pathways associated with antiseizure drug response or non-response. Herein, we not only provide a guideline for the application of these studies for screening of potential drug candidates but also suggest the other possible modifications of our approach for the development of new in vitro and etiologically relevant models to study the same. The use of these models, along with the reported ones, may open new avenues for other neuro-researchers by providing a future scope for investigation of various cell-signaling events associated with inflammation and epilepsy in the in vitro setup.

Key words Epilepsy, Neuroinflammation, Lipopolysaccharide, Pilocarpine, Seizures, HT-22 cells

1 Introduction

Inflammation is marked as a hallmark response for the cellular injury or infections that can increase the progression of a number of neuronal diseases like epilepsy and Alzheimer's disease (AD) [1, 2]. Specifically, epilepsy shares a number of pathological features associated with inflammatory responses that are substantiated by the pre-clinical as well as clinical data available [3–5]. There are also evidences delineating the effects of peripheral inflammatory conditions in epilepsy patients initiating the synthesis of inflammatory mediators in the central nervous system (CNS) [6]. A number of studies also correlate the mechanism of

Divya Vohora (ed.), *Experimental and Translational Methods to Screen Drugs Effective Against Seizures and Epilepsy*, Neuromethods, vol. 167, https://doi.org/10.1007/978-1-0716-1254-5_3,

pharmacoresistance with the inflammatory responses in epileptic animals as well as in persons with epilepsy (PWE) [7]. In few such PWE, seizures can be controlled up to a certain limit, whereas others suffer from severe, uncontrolled, or refractory (also referred as intractable or pharmacoresistant) epilepsy [8]. In 2010, a new definition of the latter was proposed by the International League Against Epilepsy (ILAE), according to which, the drug-resistant epilepsy was defined as the failure of adequate trials of two tolerated and appropriately chosen and used antiseizure drug schedules (whether as monotherapies or in combination) to achieve sustained seizure freedom [9]. Irrespective of being such a serious disease, there is still a lack of cellular models to test the existing and new chemical compounds for the treatment of inflammation-mediated seizures or to understand associated cell-signaling pathways. Though several animal models are available to study the same pre-clinically, cell-based techniques are always advantageous for the screening of the compounds designed to combat such pathological conditions. In order to overcome these limits, the development of the in vitro models to screen the anti-inflammatory compounds becomes a necessity. The development of such cellular models mimicking the actual pathophysiology of the disease ranging from the actual systemic to neuronal pathology of the PWE is still required that can be mimicked by priming the neuronal cell lines with an inflammatory trigger prior to the induction of seizures.

Also, among the several lacunae in pipeline development of molecules in the treatment of neuroinflammation, there exists a lack of proper understanding of the basic mechanisms involved in pre-clinical stage of drug discovery. This remains an arena of prime concern for the failure of several drugs in clinical trials. Several signaling pathways have been implicated in the pathophysiology of neuronal inflammation-mediated seizures, as evidenced by various immunological and pharmacological data available. The therapeutic targeting of various pathways that overactivated in neuroinflammatory pathways can be carried out using specific inhibitors, which would aid in the identification of potential molecular targets for the same. In this way, the development of in vitro models also represents a promising approach for exploration of molecular targets for the treatment of neuronal inflammation-mediated seizures in early pre-clinical stages.

In view of this, in our laboratory, we have developed an in vitro lipopolysaccharide (LPS)-primed pilocarpine (PILO) model of neuronal inflammation using mouse hippocampal cell line. This model is based on a concept of activating both inflammatory and the seizure-related pathways in neuronal cells using LPS as an inflammatory trigger, and PILO as a chemoconvulsant. In this model, we have tested the effects of three antiseizure drugs of widespread clinical use, viz., sodium valproate, carbamazepine, and levetiracetam, and observed the role of overactivated inflammatory pathways in hippocampus-like cells. In our in vivo work, the inflammatory priming (using LPS) was carried out 2 h before the pilocarpine (PILO) administration in mice, which led to the complete loss of the efficacy of valproate and carbamazepine, and a partial loss in the efficacy of levetiracetam [10]. In view of this, we also support the role of the central as well as peripheral inflammation in the mechanisms of pharmacoresistance. We could replicate these results in the in vitro situation using murine hippocampal-like cells (HT-22 cells). Based on our observations and extensive literature survey, herein, we provide details of our in vitro model for better understanding of the involvement of the neuronal inflammation pathology induced by PILO and LPS + PILO, which led to the loss in protection offered by the conventional ASDs in our in vivo model. The model is based on the same principle of inflammatory priming (by LPS) prior to the exposure of a chemoconvulsant (PILO) to the neuronal cells. Here, the cells are exposed to the LPS prior to PILO to study the effects of LPS priming in the PILO-induced neuronal cells (Fig. 1). As a modification to this method,

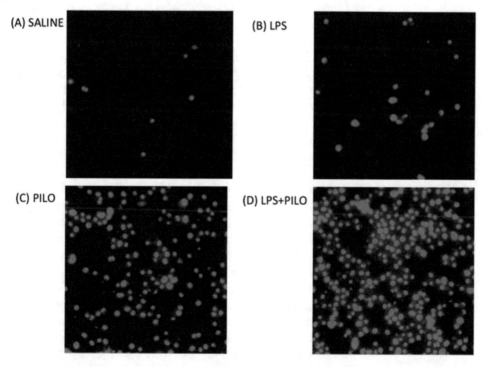

Fig. 1 The rationale of the combination model. The figure represents one of the most common mechanisms of seizure-mediated inflammatory events. Such a situation can be replicated or mimicked in the in vitro and in vivo models by priming the pilocarpine (PILO) by an inflammatory trigger like lipopolysaccharide (LPS), which is a bacterial endotoxin. LPS and PILO activate the toll-like receptor-4 (TLR-4) and muscarinic M1 receptors, respectively, leading to the activation of the inflammatory cell signaling pathways, neurodegeneration, and cell death in the brain

we also suggest the optimization of various cellular models using different neuronal cells and inflammatory mediators (other than LPS) in combination with chemoconvulsants other than PILO for the development of similar cellular models mimicking different types and extent of the disease.

2   Neuroinflammation and Epilepsy: Why Is It Important to Consider the Link?

In the past few years, various studies demonstrating the correlation between the neuronal and peripheral inflammation with the seizures have been documented [1, 3–5, 11–15], whereas the exact pathways involved in the same are still unknown. The pathological features of epilepsy also involve the release of pro-inflammatory cytokines like interleukin-1beta (IL-1β) [16], interleukin-6 (IL-6) [17, 18], and tumor necrosis factor-alpha (TNF-α) [19] that further increase the seizure susceptibility and reduce the threshold to seizures [20]. These cytokines further activate the downstream inflammatory cascades in the brain, which in turn activate the microglia and astrocytes, and impair the blood-brain barrier [21]. All these events contribute to the neuronal death as well as the structural and functional changes in the hippocampus and cortex of the brain and trigger the events associated with epileptogenesis. The neuronal death (apoptosis, autophagy, as well as necrosis) due to the development of seizures further induces the changes in the cell at a molecular level and leads to the activation of various pro-inflammatory mediators in the neuronal cell. Also, an increase in inflammatory mediators is generally observed in the cerebrospinal fluid (CSF) as well as in the serum of the PWE post-seizures [5, 22–24]. Cytokines and chemokines like IL-8 and CCL-4 are found to be elevated in the patients of pharmacoresistant epilepsy and are also associated with seizure severity [25, 26]. Other reports also associate the mechanisms of pharmacoresistance with the neuroinflammatory and hypoxic pathways [7, 27]. Therefore, neuroinflammation can be regarded both as a causative factor and a consequence of epileptic seizures.

Moving forward with the comprehension of the correlation between seizures and inflammatory events, in this chapter, we are describing a model of neuroinflammation-mediated model, which involves the activation of both inflammatory and epileptic pathways using two different inducers. It is mimicked by employing the immune-challenge models, like the systemic administration of LPS, which is a bacterial endotoxin derived from the outer cell membrane of the gram-negative bacteria [28]. LPS is known to activate the toll-like receptor-4 leading to the activation of the inflammatory pathways in the brain [29]. The systemic administration of LPS may possibly replicate the actual effects of the systemic infections on the development of seizures; however, its concomitant use with other conventional (chemical or electrical) models of seizures is reported to enhance the effects of the latter [6, 30–32]. These combination models particularly mimic the effect of systemic infections in the epileptic brain. In addition, the previous studies reported in the literature illustrate the possible involvement of the inflammatory pathways in the PILO-induced *status epilepticus* (SE), which supports its use in combination with LPS to correlate neuroinflammation-mediated epilepsy downstream of the proposed cell-signaling pathways. This particular model, as detailed in the sub-sections of this chapter, can be used to study the inflammatory pathways in neuronal cells that can be applied for the in vitro as well as the in vivo screening of the new antiseizure chemical compounds. For early pre-clinical (in vitro) pharmacological studies and high-throughput screening, assays like these can be used for studying the new targets, hit identification, and lead validation. These assays can be employed to not only study the disease mechanisms but also to narrow down the number of molecules from a series of hit and lead in drug discovery process. Utilization of such models not only augments the comprehension of the neuroinflammation-mediated pathways but also would help in probing the inflammatory mechanisms involved in the development of "pharmacoresistance" to conventional antiseizure drugs.

3   Model Features: Neuroinflammatory Pathways Induced by LPS or/and PILO

3.1   Pilocarpine (PILO) Model: Can Seizures Be Induced in a Cell Culture Dish?

Pilocarpine, a muscarinic agonist [33], is a conventional chemical method to induce seizures and chronic SE in rodent models. The pathophysiological features produced by PILO are isomorphic to the pathogenesis of temporal lobe epilepsy (TLE). It includes an initial precipitating injury from epileptic foci in the limbic system that leads to hippocampal sclerosis and neuronal death to specific brain areas like the hippocampus, cortex, neocortex amygdala, etc. [34], which is also observed as a clinical feature of the TLE in patients [35]. For induction of these events, PILO can be administered peripherally or intrahippocampally, both of which result in the development of behavioral seizures as well as the electroencephalographic (EEG) changes along with neuronal cell death. Using neuronal cell lines, these pathophysiological features can be induced using appropriate concentrations of PILO. Exposure to PILO leads to the induction of various inflammatory pathways in neuronal cells [11, 36, 37]. The cell-based pathological changes upon administration of PILO and the effect of various drugs can be tested using such models.

However, PILO exposure to cells activates the inflammatory pathways [11]; it doesn't address the issue of the existing systemic infection or any kind of existing peripheral inflammatory comorbidity in the animals. Alongside, clinically, the multiple exposures to various allergens during the lifetime result in the increase of the levels of immune mediators centrally as well as peripherally, which cannot be replicated in the inbred mice strains without any systemic trigger or priming [38]. In this vein, LPS-primed PILO models (LPS and PILO in combination) can be a better approach than the classical models of epileptogenesis to understand the contribution of the inflammation in determining the antiseizure potential of the therapeutic compounds.

3.2 LPS + PILO Model (Combination): How LPS Priming Can Mimic the Actual Pathophysiological Pattern of Neuronal Inflammation-Mediated Epilepsy?

The administration of LPS is one of the most common models to induce neuronal inflammation in animals. It activates the microglia via the toll-like receptor-4 (TLR4) pathway and further leads to the increased release of the pro-inflammatory mediators involved in seizure pathophysiology [21, 39, 40]. The in vivo data available demonstrates different effects of LPS administration in various models of epilepsy based on its time of administration. It has shown to protect against seizures if administered 18–24 h prior to PILO-induced seizures; it shows an anti-convulsant effect [41, 42] and behaves as a pro-convulsant when administered 4 h prior to pentylenetetrazole exposure. Also, it has shown protection against the cell death when exposed to the cells 72 h prior to seizure induction. Controversial reports have also suggested no effect of LPS priming prior to seizures [43]; however, when administered 1–2 h before seizures, it behaves as a pro-convulsant [32, 44]. In correlation with these, our model is based on the fact that the priming with LPS prior to PILO in cell-based models can change the excitatory changes in the membrane of the neuronal cell and increase the severity along with neuronal degeneration (as compared to induction with PILO alone) [45].

4   Materials Required

4.1 Basic Reagents

Both PILO and LPS (from *Escherichia coli*) can be procured commercially. As observed by us, both PILO and LPS can be dissolved in 0.9% NaCl solution or Milli-Q water; however, storing them for more than 24 h is not recommended. Also, caution should be taken, and LPS should be stored in an amber-colored container as it is light sensitive. If stock solution for LPS has to be prepared, considering its low dose, it should be prepared in DMSO and should be further diluted in Milli-Q water. Another observation that we have made in our laboratory work is that the stock solution of LPS should be vigorously vortexed before dilution, as it tends to settle on the walls of the micro-centrifuge tubes. For cell culture experiments, the solutions should be prepared in the autoclaved water and should be filtered using the bacteriostatic filters of the pore size of 0.22μ. All other reagents used in the experiments were of analytical grade. Milli-Q water was used throughout the experimental work.

4.2 Cell Lines and Cell Culture Requirements

The HT-22 cells (sub-clone of HT4 mouse hippocampal cell line) are cultured and grown to 80–90% confluency in DMEM-high glucose media supplemented with 10% heat-inactivated Fetal Bovine Serum (FBS) and 1% antibiotic-antimycotic solution. Though our laboratory research was restricted to the mouse hippocampal-like cells (HT-22), we propose that the human neuroblastoma-like cells (SH-SY5Y cell line), rat hippocampus-like cells, and albino mouse neuroblastoma cells (Neuro 2A) can also be explored. All the neuronal cultures should be grown by incubating at 37 °C with 5% CO_2 using CO_2 incubator. For the conduct of this in vitro work, basic cell culture facility and aseptic conditions are required.

4.3 Other Chemicals Required

At the time of sample preparation, total protein should be calculated using the Bradford reagent test. For the Bradford test, bovine serum albumin (BSA) is required as the standard. Also, Bradford reagent and dimethylsulfoxide (DMSO) are required for the test procedure. Other methods like for total protein estimations, Bronsted-Lowry method, can also be used.

Requirement of other chemicals is dependent upon the type of assay the experimenter wants to conduct. Assessment of different parameters after sample preparation affects the assay requirements accordingly.

5 Methods and Considerations

5.1 Cell Culturing

In the process of cell culturing, the cells are grown under controlled conditions. They can either be attached to the surface of the flask (adherent cultures) or can be present unattached in suspension form (suspension culture). These cells can be derived directly from an organism and are known as the *primary cultures*. These cells can be initially fractionated to separate different cell types or can be cultured as such. These types of cells can also be sub-cultured for weeks and can be allowed to proliferate in new flasks repeatedly. However, these cells have a particular life span, and cannot be sub-cultured after a specific number of passages due to *cell senescence*. These cells can be immortalized by spontaneous mutations or genetic manipulation and can be used indefinitely in the form of cell lines. For passaging (sub-culturing or splitting) of these cells, a small number of cells are transferred into fresh media in a new culture flask as mentioned in subsequent sub-sections of methodology. These cell lines can be cultured for longer period of time if passaged regularly to avoid the cell senescence and death associated with the over-confluency of the culture flasks. Alongside, the new freeze downs can be prepared from the healthy stock and can be stored in liquid nitrogen for later usage (*see* **Notes 8.1 and 8.2**).

The following basic steps are involved in culturing of HT-22 cells that we used in our study. This method can be used to culture any mammalian cell line after optimization of the procedure as per the cell type and origin.

5.1.1 Media Preparation

1. The liquid DMEM media can be procured from a commercial source, which can be used directly after addition of FBS and antibiotic solution.

2. If not available, powdered media can be procured, which can be suspended in the autoclaved Milli-Q water. Ten percent FBS and $1\times$ Penicillin-Streptomycin (antibiotic solution) are also added to the media prepared. The required amount of sodium bicarbonate is also added to the prepared media, if mentioned on the commercially available DMEM media sachet.

5.1.2 Thawing and Reviving Frozen Cell Lines

The thawing procedure is generally stressful to frozen cells (specially to neuronal cells as they are oversensitive to the change in environmental conditions). Appropriate use and good cell culture practices and procedures help in the survival of a higher number of cells.

1. Frozen cryo-vials should be carefully removed from liquid nitrogen tank, wiped with alcohol and immediately transferred to 37 °C incubator.

2. The 5 mL of pre-warmed media is quickly transferred into T-25 culture flask.

3. When completely thawed, the cells were transferred into the 5–7 mL of media (depending upon the requirement) in a centrifuge tube and centrifuged for 5 min at 1000 rpm. The g-force or the relative centrifugal force (RCF) exerted here will also depend on the radius of the rotor (cm) of the centrifuge used in the laboratory. For this purpose, RCF can be calculated as $(RPM)2 \times 1.118 \times 10^{-5} \times r$, where r is the centrifuge radius in cm. The pellet obtained is re-suspended in fresh media and is transferred to the T-25 flask (as mentioned above).

4. The flasks are then incubated at 37 °C, 5% CO_2 incubator.

5.1.3 Maintenance of Mammalian Cell Line In Vitro Culture

1. On confluency, all the media are removed from the T-25 flask with a sterile micropipette, and cells are washed using sterile $1\times$ PBS.

2. The cells were detached using cell scraper or trypsin (using general protocol), and the cell suspension is centrifuged at 1000 rpm, for 5 min.

3. The cell pellet obtained is again washed with $1\times$ PBS and subsequently re-suspended in the fresh media.

4. To split cultures, 1/3rd or 1/4th of this cell suspension is added in the sufficient medium (as prepared in **step 1**) such

that 500μL of the cell suspension can be transferred into each fresh culture flask.

5. Alternatively, the cells are counted using a cell counter or a hemocytometer before plating.

5.1.4 Cell Counting Using Hemocytometer

1. 0.4% solution of trypan blue is used to stain the cells in the ratio of 1:1 or 1:10 according to the cell density.

2. 10μL of cell suspension is loaded in a hemocytometer, and cell viability along with the cell density was examined under the microscope at low magnification.

3. The blue-stained cells and the total number of total cells. The cells that take up trypan blue are considered non-viable.

4. The suspension of healthy cells with cell viability of around 90–95% should be used for further experiments. Counted cells are plated in the petri plates or the well-plates as per the requirement of the subsequent experimental procedure. All log phase cultures that we used while conducting our experiments were healthy.

5.2 Optimization of the Cellular Model

1. For development of a cellular model for any in vitro experiment, firstly, it is always required to optimize the model by optimizing different variables like cell density, concentrations of inducers and test compounds, incubation time, media composition, etc. Several controls need to be considered including vehicle control, positive controls (comparison group), negative control (induced-untreated cells), per se group (un-induced-treated), and wells containing no cells (Fig. 2).

2. All the experiments should be reproducible, and condition controlled. Z should be considered as the measure of assay quality. A higher Z value conveys greater reproducibility and precision of the assay. A perfect assay has a Z value of 1. For a high-throughput screening assay, a Z value of >0.5 is generally considered acceptable.

3. In this particular model, for the optimization purpose, the murine hippocampal cells represented by the HT22 cells are plated at an optimized cell density in the complete DMEM high glucose media (containing 10% heat-inactivated FBS and 1% antibiotic).

4. It should be noted that in natural environment, the hippocampal cells usually cross talk with the extracellular matrix and other cells in the brain that may affect the regulation of various neuroinflammatory pathways. In the pure cell line culture, these interactions between various cell types are either lost or reduced, and therefore, cannot be studied well. The optimization of the cell culture media and culturing the cells in the presence of the extracellular components including various

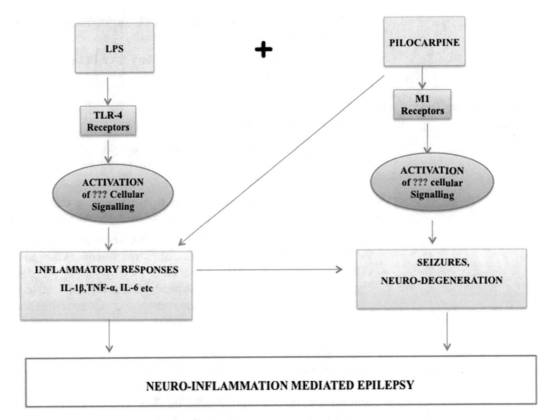

Fig. 2 Assay variables for the optimization of the cellular model. Herein, it is shown that each and every parameter should be optimized before carrying out the in vitro studies. It is shown that a range of the number of cells per well should be tested with various concentrations of the inducers and the test compounds at different time points of incubation. Other assay variables as mentioned in the subsections of the chapter can also be considered depending upon the type of assay conducted

proteins and salts would be a better representation of the natural environment. Such 3D models are more predictive of the actual pathophysiology of the disease and can be used to study neurogenesis, neuronal damage, migration, and apoptosis in a better way [46].

5.3 Cell Plating, Sample Preparation, and Analysis of Biomarkers

1. For analysis of various biomarkers, unless any specific requirement is there, the adherent HT22 cells are grown in the cell culture plates using high glucose-DMEM media (containing 10% heat-inactivated FBS and 1% antibiotic).

2. In our work, we have used both 96-well plates and 100 mm petri dishes for these experiments. Cell densities to be plated in these plates were individually optimized for different experiments.

3. Cells are plated and incubated for 24 h so that they grow and gain their actual morphology after passaging. Subsequently, the cells are exposed to the test compounds followed by simulation

with the optimized concentration of LPS after 30 min and/or PILO after 1 h subsequent to drug treatments.

4. After optimized time duration subsequent to the exposure, the cell supernatants are collected, and the samples were prepared using 0.5 M phosphate buffer, 1% NP-40 buffer, or the RIPA buffer (150 mM sodium chloride, 1.0% NP-40, 0.5% sodium deoxycholate, 0.1% SDS; sodium dodecyl sulfate, 50 mM Tris-HCl, pH 8.0) supplemented with protease inhibitor cocktail and PMSF (phenylmethylsulfonyl chloride, a serine protease inhibitor).

5. This lysate can be used for analysis of various markers of inflammation and death-like cytokines, proteins, and other mediators using ELISA or western immunoblotting techniques. Other experiments can also be used for cell imaging, fluorescence microscopy (FM), RT-PCR, and flow cytometry, and comparative analysis can be carried out.

6. These techniques can also give an idea about the morphological changes in the dying cells and the pattern of cell death subsequent to seizures. Here, in Fig. 3, we have stained the cells with propidium iodide (PI) after fixation (in 4% PFA) and stabilization. In the figure, it is clearly seen that the saline group shows the least damage (least number of stained cells); however, the LPS group showed slight disruption of nuclear integrity. In the PILO group, increased damage is observed upon PI staining, which was further potentiated in the combination group.

Assay variables (in vitro optimization):

Number of cells/ well	concentration	Incubation Time

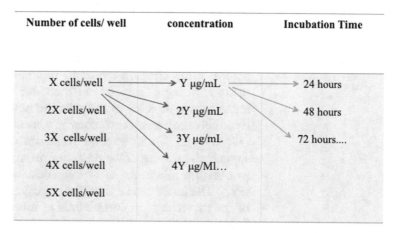

Fig. 3 Fluorescence microscope images (60×) of propidium iodide stained HT-22 cell line posttreatment with LPS (1 μg/mL), PILO (200 μM), and LPS + PILO (1 μg/mL + 200 μM). Here, saline group represents the vehicle treated group with 0.9% NaCl

7. In a similar way, neuronal electrophysiological studies can also be carried out in these cells post-induction with LPS and/or PILO; however, the concentrations of different ions involved in neuronal electrophysiology (like sodium and potassium) should be mentioned in the media. At rest, the sodium channels on the neuronal membrane are closed, and a number of potassium channels are open that maintains the resting membrane potential. This changes upon induction with the inducers that increase the neuronal excitability, which is detailed in other sections of this book. Similar studies can be conducted using neuronal cell lines, and this membrane potential can be measured [47].

6   Observations and Interpretation of Data

6.1   Why This Particular Model?

There are two types of studies that can be carried out using this model. The *first* one involves the corroboration of the molecular mechanisms involved in the pathogenesis of the neuronal inflammation-mediated seizures. The inhibitors of downstream mediators involved in different cell signaling pathways activated by LPS and/or PILO can be used in this model for the better understanding of the cell-signaling events. In our work, we conducted various experiments to understand the pattern of neuronal inflammation and cell death post LPS and/or PILO exposure to the HT-22 cells. We observed an increase in the oxidative load, reactive oxygen species levels, and increased cell death in PILO groups that was further potentiated upon the LPS-priming. Considering the same, we suggest this model as an appropriate method to measure the levels and expressions of different mediators involved in various inflammatory pathways. In this way, it can be used to decipher the involvement of possible mechanisms of seizure and neuronal inflammation-associated cell death.

The *second* one is that such in vitro models of neuronal inflammation using chemoconvulsants in combination with the inflammatory priming can be further used to evaluate the anti-inflammatory efficacy of series of the new chemical entities meant to inhibit the certain molecular targets and thus be more etiologically relevant. In an attempt to screen new compounds, we also suggest the testing of varying concentrations of inhibitors ranging to obtain a minimum of 6-point dose-response curve (for precision). The latter can be used for the calculation of the IC50 values of the new chemical compounds in murine as well as human cell lines.

6.2   What We Observed in Our Experiments?

For the elucidation of the mechanism of neuronal cell death associated with inflammation in neuronal cells, we first calculated the IC50 of valproate, carbamazepine, and levetiracetam using the in vitro model (optimization) and used the screened concentrations of these drugs for further experiments. In our further experiments, we observed a dramatic elevation in the cell death and oxidative burden upon LPS priming in the PILO model. We also tried to understand the mechanism of this cell death associated with PILO as well as LPS + PILO and studied the biomarkers for apoptotic and necrotic cell death, respectively. The increase in the levels of caspase-3 in LPS, PILO, and LPS + PILO groups confirmed the involvement of apoptotic cell death. We also found increased LDH leakage upon PILO as well as LPS + PILO exposure indicating the presence of both apoptotic and necrotic cell death; however, the data revealed the mechanism to be predominately apoptotic. Sodium valproate and carbamazepine, but not levetiracetam, decreased the caspase-3 cleavage in LPS + PILO model, which was accompanied with reduced lactate dehydrogenase (LDH) leakage in the PILO model but failed to produce any significant decrease in the LPS + PILO model. We showed the involvement of necrotic pathways in neuroinflammation-mediated seizures, but our results also ruled out the possibility of it being the major pathway involved in the development of resistance. On the other hand, the higher expression of caspase-3, which was not attenuated by the antiseizure drugs, might be an indication of the predominant involvement of a pro-apoptotic but not the necrotic signaling in the process of loss of protection by the conventional drugs [10]. This model demonstrated the contribution of neuro-inflammatory pathways in the development of pharmacoresistance (loss in protection) with the conventional antiseizure drugs in the neuronal cells. Apart from the abovementioned, other inflammatory mediators and electrophysiological changes can also be studied in this model to further understand the involvement of mechanistic pathways, and screening of new antiseizure compounds using the same.

6.3 LPS- and/or PILO-Induced Cell Death: Is There Any Possibility of False-Positive Results if This Is Tried In Vivo?

The neuronal damage and cell death subsequent to PILO-induced seizures is time specific for different areas of the brain. It begins within first few hours of induction and progresses up to months. In the in vivo models, the rodents exposed to PILO show the signs of cell death within few hours of induction, which mainly involve the superficial layers of the cortical areas [48]. With time, this damage is further increased, and other areas also start getting affected (for details, please refer to Chapter 7). In case of cell lines, these events including morphological changes and pattern of cell death can be studied by assessment of various parameters associated with cell death and proliferation. Various assays using the water-soluble or insoluble tetrazolium dyes like -(4,5-dimethylthiazol-2-yl)-2,5-diphenyl tetrazolium bromide (MTT), 3-(4,5-dimethylthiazol-2-yl)-5-(3-carboxymethoxyphenyl)-2-(4-sulfophenyl)-2H-tetrazolium (MTS), and 2-(2-methoxy-4-nitrophenyl)-3-(4-nitrophenyl)-5-(2,4-disulfophenyl)-2H-tetrazolium (CCK-8 or WST-8) can be carried out [49]. Also, there are other assays where lysis of the cells is not required for the collection of samples and, therefore, are ideal for the kinetic studies using cell culture experiments like resazurin (7-hydroxy-3H-phenoxazin-3-one 10-oxide) [50] and XTT (2,3-bis [2-methoxy-4-nitro-5-sulfophenyl)-5-((phenylamino) carbonyl]-2H-tetrazolium hydroxide) [51], which do not require cell lysis and which allow kinetic monitoring. Using these dyes, samples can be collected at different time points for the measurements of various biomarkers in the treated or untreated cells (Also, *see* **Note 8.3**).

Other associated changes like change in nuclear integrity, DNA fragmentation, LDH leakage from the cells, intracellular calcium levels, and ATP/ADP ratios in populations of the cultured neuron-line cells can be studied. Pacico and Meur [52] have also presented an assay for the fluorescent measurements of the synchronized neuronal calcium oscillations in primary neuronal cultures [52]. Flow cytometric analysis of cell-cycle and associated events can be carried out subsequent to propidium iodide (PI) staining to understand the effects of various chemical compounds on the cell death induced by LPS and/or PILO. Further, the expressions of protein mediators involved in the cell death and apoptosis and necrosis can be studied. The involvement of various cell-signaling pathways can also be deciphered by measurement of the protein expressions involved in those pathways by western immunoblotting techniques.

In the in vivo setup, the administration of both LPS and PILO is known to produce the neuronal degeneration in rodents post induction in a time-dependent fashion [35, 53]. Specifically, PILO administration leads to the degeneration of the inter-neuronal cells in the cornus ammoni (CA) subareas of the brain including CA1 and CA3 [54]. Apart from that, severe neuronal damage is observed in the dentate gyrus region of the hippocampus in addition to the mild effect in other areas of the brain like the amygdala, thalamus, cortex and neocortex, and substantia nigra [48, 55]. Also, in the samples obtained from PWE during the surgical procedures, hippocampal sclerosis (most common in adults), ganglioglioma tumors, and malformations of cortical development (mostly in children) are reported as the most common histopathological changes associated with the disease [53]. Therefore, studying the histopathological changes post LPS + PILO-induced seizures is a mandate in this model as well (*see* **Note 8.4**).

Additionally, we also suggest the experimenter to consider the excessive proliferation and neurogenesis of the neuronal cells subsequent to the intermittent as well as the prolonged seizures in the rodent models [56–58]. There are evidences that the sub-granular cells in the DG region of hippocampus show increased proliferation just after the development of SE; however, these cells are not able to survive for more than 2–3 weeks post-SE, and the rate of proliferation bounce back to the baseline [59, 60]. Due to the same reason, it is recommended to also observe the long-term effects of the drugs on neurodegeneration in this LPS + PILO model to avoid the false-positive results in the drug treatment groups. It can be due to the excessive proliferation of the hippocampal cells post-SE rather instead of the true pharmacological effect of the drug. Different stains like cresyl violet, Fluoro-Jade, etc., can be used to study neuronal pyknosis and death [37]. Also, this pattern of post-SE neurogenesis in LPS + PILO model can be validated in the in vitro studies using the cell lines like SH-SY5Y [61].

7   Use of Other Cell Lines and Inducers: Is There Any Translational Value?

In addition to the use of rodent cell lines like HT-22, rat hippocampus-like cells [62], and albino mouse neuroblastoma cells (Neuro 2A) [63], human cell lines can also be used to study the pathological features from the clinical perspective. For this purpose, cell lines like human neuroblastoma-like cells (SH-SY5Y cell line) [64] and brain microvascular endothelial cell line hCMEC/D3 (human blood-brain barrier-like cells) [65] can be employed to study the effects on neuronal death, neurogenesis, and blood-brain barrier disruption in case of neuronal inflammation-mediated seizures or epilepsy.

For blood-brain barrier studies, the primary cerebral microvessel endothelial cells would always remain a better model, but it is impractical to always work with the human cerebral tissues to such a wide extent in laboratory conditions. Due to this fact, hCMEC/D3 cells can be tried to perform all the cellular and molecular studies to study the effects of various drug molecules and their transport mechanisms in these models in correlation with inflammation-mediated epilepsy models [65]. Human-derived induced pluripotent stem cells (iPSCs) are also evidenced to be used for the neuro-toxicological testing of the drug molecules [66]. These days, the multi-electrode systems are also being used with these in vitro cultures to study the changes in the electrical activities, and the epileptiform bursts as well as cell signaling pathways [67–69]. Potentiation of seizure-like activities in cell-based models can be studied at different time points using two or more inducers as an extension to our study.

There are evidences for the use of kainic acid [70] and PTZ in combination with an inflammatory trigger to induce the neuroinflammation-mediated seizures in the in vivo models. LPS priming 2–2.5 h prior to the induction of hypothermic seizures induced by hair dryer or heat lamp in rat and mice pups also showed the potentiation of seizure-induced pro-inflammatory cytokine production and microglial activation [71]. Similarly, in the in vitro systems, various inducers can be tried and optimized to see if they produce the similar neuroinflammation-mediated epilepsy like pathology or not. Notably, kainic acid has already been studied to produce the epilepsy-like excitotoxicity and cell death in the in vitro systems [72, 73]. PTZ has also shown to produce apoptotic cell death in neuronal cultures [74] and can be explored in combination in the in vitro models. This way, more models would be developed for the screening of the new antiseizure drugs. In addition to LPS, we recommend other inflammatory mediators like IFN-γ, TNF-α, IL-8, etc., which can also be tried as per the study requirement and design as provocative or protective stimuli at different time points in combination with a convulsant [75, 76].

8   Notes and Troubleshooting

Though most of the troubleshooting methods and the basic considerations are mentioned throughout the chapter, the remaining and the most common ones are detailed below.

8.1   Basic Cell Culture Issues

1. For cell-based studies, the cellular morphology and growth rate of the neuronal cells should be checked regularly before using the same for experiments. The appropriate culture conditions should be provided to maintain the neuronal health. Passaging should be carried out at regular intervals to promote the neuronal growth. Edge effects should be avoided by maintaining the uniform cell density in the culture flask as well as the experimental plate.

2. Equal density of the cells should be plated in each well/plate to avoid the false results.

3. Promote consistent cell density and health throughout your culture vessels by avoiding edge effects. Cell suspensions should be mixed well and should not be vigorously vortexed or pipetted for longer duration. Occasional pipetting should be carried out for uniform plating of the neuronal cells.

4. After plating your neurons, the plates should always be observed under the microscope to assure the uniform distribution of cells.

5. Poor attachment of cells can be due to bacterial contamination, incorrect freezing, storage, or thawing of stock. Poor quality of consumables (like cell culture media) or harvesting the cells from over-confluent cultures can also lead to poor attachment of cells in the new culture flasks. In such cases, new stock of cells and media should be tried. Also, appropriate media meant for the particular neuronal cells should be used to avoid such cases. Once plated, cells should not be disturbed by any kind of

shaking to avoid any variations in the cultures. Cells should be allowed to attach without disturbing as much as possible.

8.2 Problem of Contamination

Obtaining cell lines' authentic sources, characterization of the cells upon receipt of cell line (and at particular interval of time), and good aseptic technique help in avoiding the problems associated with the contamination of the cells. Few precautions required to avoid the same are mentioned in the subsequent sections.

1. The cell culture hood (biosafety cabinet) used to carry out the cell culture work should be located in an area that is strictly restricted to cell culture practices.

2. The surface should be uncluttered and contain only items required for a particular procedure. No extra amount of material should be kept inside the hood.

3. The surface of the hood should be disinfected with 70% alcohol before and after every usage apart from the routine cleaning of the equipment and the culture room. In case of spillage, it should be wiped immediately using 70% alcohol to avoid any chances of contamination.

4. Ultraviolet light can also be used to sterilize exposed work surfaces in the cell culture hood between uses.

5. Hands should be washed properly before and after working with cell cultures reagents. Also, gloves and mask should be worn to reduce the possibility of contamination from skin and clothes. All the reagents and consumables used in the cell culture experiments, irrespective of the quality, can be contaminated while handling.

6. Commercial reagents and media undergo strict quality control to ensure their sterility, but they can become contaminated while handling. Follow the guidelines below for sterile handling to avoid contaminating them. Always sterilize any reagents, media, or solutions prepared in the laboratory using the appropriate sterilization procedure (e.g., autoclave, sterile filter).

7. Always wipe your hands and your work area with 70% ethanol.

8. Wipe all the containers, flasks, plates, and dishes with 70% ethanol before placing them in the cell culture hood. Also, empty glass bottles should be autoclaved separately before using them in culture hood.

9. Pipettor should be used with the disposable pipettes for liquids. The pipettes should not be reused to avoid cross contamination.

10. Serological pipettes and pipette tips should be used carefully and should not be reused. The pipette tips should not be touched anywhere unnecessarily to avoid contamination.

11. All the packets of the consumables should only be opened in the culture hood and should never be left open outside.

12. Do not remove the lid of any flask, bottle, or petri dish until it is required to be used. It should never be left open in the environment without purpose and should be covered back as quickly as possible. The lids of the bottles and petri dishes should never be kept on the surface facing down. The rim of the lid should not be touched.

13. Only sterile glassware should be used for cell culture experiments.

8.3 Other Considerations

1. False positives or erroneous results can be obtained if the samples are not prepared with proper care and skill. The volumes as well as the total protein content in the samples (to be loaded for biochemical estimations, western blots or ELISA) should be equalized by the total protein analysis using the Bradford reagent test. In this, several dilutions of the samples obtained from the neuronal cultures or the animal experiments are prepared.

2. 1 mg/mL stock of bovine serum albumin (BSA) is also prepared as a standard. Diluted samples as well as the standards are added to uncoated ELISA plates, and subsequently, Bradford reagent is added to each well. Absorbance is measured at 595 nm after 5 min to 1 h of incubation (after development of the light purplish-blue color) [77].

3. Total protein concentrations are extrapolated with the help of the standard curve. Subsequently, for any analysis, equal total protein concentrations should be loaded for each sample to avoid the false positives. Volumes of the samples should also be equalized using the diluent.

8.4 For Animal In Vivo Studies in Rodents (Using the Similar Concept of Inflammatory Priming)

1. Pre-treatment with LPS should be carried out 2 h before the PILO induction. Variation in the time points of administration can lead to variable results.

2. Before the administration of PILO, an anti-cholinergic agent should be administered to limit the peripheral cholinergic side effects of PILO [35].

3. During behavioral seizure scoring care must be taken to place single animal per cage to avoid animals hurting each other in severe seizure conditions.

4. Post-PILO treatment, especially if primed with LPS, leads to the deterioration in animal health. Some of those animals are not able to move or consume food just after the seizure termination and diazepam injection. In such situation, in our own study, we attempted to improve the survival rate in the animals by providing 5% dextrose solution in a separate container in addition to normal water, which is usually provided.

5. Anesthetic agents, if used, should minimally interfere with the study results. They should be selected after extensive literature search to minimize interruptions with the outcomes. Generally, isoflurane is used an anesthetic of choice; however, a combination of ketamine and xylazine can also be administered.

6. Animals should be carefully monitored everyday post-SE, and their health and recovery should be assured before conduct of any other behavioral tests for the neuropsychiatric manifestations associated with epilepsy.

7. Hypothermic response is generally observed in rodents upon LPS administration [78], and therefore, the rectal temperature of the animals can be measured periodically to assure good health.

8. Considering the pathological effects of the combination model, the experimenter can consider the calculation of the organ-body weight index (OBI) for each animal. For this purpose, each organ (brain, heart, kidney, lungs, liver, pancreas, etc.) is excised at the time of animal sacrifice and quickly rinsed in 0.9% saline, blotted, and weighed [79].

$$OBI = (Organ\ weight/animal\ weight) \times 100.$$

9   Conclusion

After 50 years of the discovery and establishment of the conventional antiepileptic drugs, the need for the evaluation of new compound hitting newer targets has emerged. However, among several lacunae in the development of these molecules, there exists a lack of proper understanding of the cell-signaling pathways in the pre-clinical stages of the drug development. As estimated by ILAE, the emerging pharmacoresistance from the conventional antiseizure drugs has become the prime concern in the present arena, which has additionally obligated the utilization of novel pharmacological models for the investigation of different molecular targets and associated cell-signaling pathways. In this scenario, the foremost challenge is the identification and the selection of innovative targets for treating the same. In this view, we tried to probe the mechanisms of resistance to the antiseizure drugs using neuroinflammation-mediated in vitro and in vivo models. Our in vivo work demonstrated the involvement of inflammatory pathways in the cell signaling mechanisms responsible for the development of resistance (loss of protection) from conventional antiseizure drugs upon LPS-priming in PILO model of seizures. Our in vitro model based on the same principle of inflammatory priming (as detailed in this chapter) to study the mechanisms of pharmacoresistance associated with the same. Herein, we studied the inflammatory pathways that are activated upon induction with a conventional chemoconvulsant (PILO). Alongside, we primed this PILO model with an inflammatory trigger (LPS) to check how these pathways are activated in neuronal cells upon exposure with conventional chemoconvulsant combined with inflammatory trigger. To work in this direction, we first optimized the model using HT-22 cells and evaluated various concentrations of inducers and test drugs at different time points. In our further experiments, we observed a dramatic elevation in the cell death and oxidative burden upon LPS priming in the PILO model. We tried to understand the mechanism of this cell death associated with PILO as well as LPS + PILO and studied the biomarkers for apoptotic and necrotic cell death, respectively. We showed the involvement of necrotic pathways in neuroinflammation-mediated seizures, but our results also ruled out the possibility of it being the major pathway involved in the development of pharmacoresistance. The higher expression of caspase-3, which is a marker of apoptotic cell death, was not reduced by the antiseizure drugs. This indicates the major involvement of pro-apoptotic but not the necrotic pathways in the development of pharmacoresistance [10].

In addition to our work, other studies also present a strong evidence of the correlation of neuronal inflammation and seizures, and the involved cell signaling pathways and mediators can serve as attractive drug targets for seizures. In this case, the model discussed in this chapter can be used for the exploration of the new chemical compounds targeting the neuroinflammatory pathways associated with epilepsy. Use of such models (in addition to in vivo studies) would not only augment the current comprehension of the participation of these pathways in the progression of epilepsy but would also probe their contribution in the mechanisms involved in "pharmacoresistance." Added insights into this area may prove to be of great significance for better understanding of the inflammatory pathways as the potential targets of exploration for the treatment of the same. This model, along with the reported ones, can be used for the screening on antiseizure molecules in early pre-clinical stages of drug development, which can open new doors for other neuro-researchers by providing a future scope for further investigation on various cell-signaling events associated with inflammation and epilepsy.

Acknowledgment

The laboratory of Prof. Divya Vohora is funded by the Department of Science and Technology under the DST-FIST program and the All India Council for Technical Education under the AICTE-MODROBS program of the Department of Pharmacology, School of Pharmaceutical Education and Research (SPER), Jamia Hamdard.

References

1. Alyu F, Dikmen M (2017) Inflammatory aspects of epileptogenesis: contribution of molecular inflammatory mechanisms. Acta Neuropsychiatr 29(1):1–16. https://doi.org/10.1017/neu.2016.47

2. Kinney JW, Bemiller SM, Murtishaw AS, Leisgang AM, Salazar AM, Lamb BT (2018) Inflammation as a central mechanism in Alzheimer's disease. Alzheimers Dement (NY) 4:575–590. https://doi.org/10.1016/j.trci.2018.06.014

3. Kleen JK, Holmes GL (2008) Brain inflammation initiates seizures. Nat Med 14 (12):1309–1310. https://doi.org/10.1038/nm1208-1309

4. Vezzani A, Friedman A (2011) Brain inflammation as a biomarker in epilepsy. Biomark Med 5 (5):607–614. https://doi.org/10.2217/bmm.11.61

5. Aronica E, Crino PB (2011) Inflammation in epilepsy: clinical observations. Epilepsia 52 (s3):26–32. https://doi.org/10.1111/j.1528-1167.2011.03033.x

6. Riazi K, Galic MA, Pittman QJ (2010) Contributions of peripheral inflammation to seizure susceptibility: cytokines and brain excitability. Epilepsy Res 89(1):34–42. https://doi.org/10.1016/j.eplepsyres.2009.09.004

7. Pedre LL, Chacón LMM, Orozco-Suárez S, Rocha L (2013) Pharmacoresistant epilepsy and immune system. In: Pharmacoresistance in epilepsy. Springer, New York, pp 149–168

8. Beghi E (2020) The epidemiology of epilepsy. Neuroepidemiology 54(2):185–191. https://doi.org/10.1159/000503831

9. Kwan P, Arzimanoglou A, Berg AT, Brodie MJ, Allen Hauser W, Mathern G, Moshé SL, Perucca E, Wiebe S, French J (2010) Definition of drug resistant epilepsy: consensus proposal by the ad hoc Task Force of the ILAE Commission on Therapeutic Strategies. Epilepsia 51(6):1069–1077. https://doi.org/10.1111/j.1528-1167.2009.02397.x

10. Vyas P, Kumar Tulsawani R, Vohora D (2020) Loss of protection by antiepileptic drugs in lipopolysaccharide-primed pilocarpine-induced status epilepticus is mediated via inflammatory signalling. Neuroscience 442:1. https://doi.org/10.1016/j.neuroscience.2020.06.024

11. Ravizza T, Gagliardi B, Noé F, Boer K, Aronica E, Vezzani A (2008) Innate and adaptive immunity during epileptogenesis and spontaneous seizures: evidence from experimental models and human temporal lobe epilepsy. Neurobiol Dis 29(1):142–160. https://doi.org/10.1016/j.nbd.2007.08.012

12. Devinsky O, Vezzani A, O'Brien TJ, Jette N, Scheffer IE, de Curtis M, Perucca P (2018) Epilepsy. Nat Rev Dis Primers 4:18024. https://doi.org/10.1038/nrdp.2018.24

13. Vezzani A, Friedman A, Dingledine RJ (2013) The role of inflammation in epileptogenesis. Neuropharmacology 69:16–24. https://doi.org/10.1016/j.neuropharm.2012.04.004

14. Vezzani A, French J, Bartfai T, Baram TZ (2011) The role of inflammation in epilepsy. Nat Rev Neurol 7(1):31–40. https://doi.org/10.1038/nrneurol.2010.178

15. Vitaliti G, Pavone P, Mahmood F, Nunnari G, Falsaperla R (2014) Targeting inflammation as a therapeutic strategy for drug-resistant epilepsies. Hum Vaccin Immunother 10 (4):868–875. https://doi.org/10.4161/hv.28400

16. Peltola J, Palmio J, Korhonen L, Suhonen J, Miettinen A, Hurme M, Lindholm D, Keränen T (2000) Interleukin-6 and interleukin-1 receptor antagonist in cerebrospinal fluid from patients with recent tonic–clonic seizures. Epilepsy Res 41(3):205–211

17. Lehtimäki KA, Keränen T, Huhtala H, Hurme M, Ollikainen J, Honkaniemi J, Palmio J, Peltola J (2004) Regulation of IL-6 system in cerebrospinal fluid and serum compartments by seizures: the effect of seizure type and duration. J Neuroimmunol 152 (1):121–125. https://doi.org/10.1016/j.jneuroim.2004.01.024

18. Peltola J, Hurme M, Miettinen A, Keränen T (1998) Elevated levels of interleukin-6 may occur in cerebrospinal fluid from patients with recent epileptic seizures. Epilepsy Res 31 (2):129–133

19. Patel DC, Wallis G, Dahle EJ, McElroy PB, Thomson KE, Tesi RJ, Szymkowski DE, West PJ, Smeal RM, Patel M, Fujinami RS, White HS, Wilcox KS (2017) Hippocampal TNFα signaling contributes to seizure generation in an infection-induced mouse model of limbic epilepsy. eNeuro 4(2):ENEURO.0105–0117.2017. https://doi.org/10.1523/ENEURO.0105-17.2017

20. Rao RS, Medhi B, Saikia UN, Arora SK, Toor JS, Khanduja KL, Pandhi P (2008) Experimentally induced various inflammatory models and seizure: understanding the role of cytokine in rat. Eur Neuropsychopharmacol 18

(10):760–767. https://doi.org/10.1016/j.euroneuro.2008.06.008

21. Marchi N, Granata T, Ghosh C, Janigro D (2012) Blood-brain barrier dysfunction and epilepsy: pathophysiologic role and therapeutic approaches. Epilepsia 53(11):1877–1886. https://doi.org/10.1111/j.1528-1167.2012.03637.x

22. Lehtimäki KA, Keränen T, Palmio J, Mäkinen R, Hurme M, Honkaniemi J, Peltola J (2007) Increased plasma levels of cytokines after seizures in localization-related epilepsy. Acta Neurol Scand 116(4):226–230

23. Lee VLL, Shaikh MF (2019) Inflammation: cause or consequence of epilepsy? In: Epilepsy-advances in diagnosis and therapy. IntechOpen, London

24. Yu N, Di Q, Hu Y, Zhang Y-f, Su L-y, Liu X-h, Li L-c (2012) A meta-analysis of pro-inflammatory cytokines in the plasma of epileptic patients with recent seizure. Neurosci Lett 514(1):110–115. https://doi.org/10.1016/j.neulet.2012.02.070

25. Youn Y, Sung IK, Lee IG (2013) The role of cytokines in seizures: interleukin (IL)-1β, IL-1Ra, IL-8, and IL-10. Korean J Pediatr 56 (7):271

26. Cerri C, Genovesi S, Allegra M, Pistillo F, Püntener U, Guglielmotti A, Perry VH, Bozzi Y, Caleo M (2016) The chemokine CCL2 mediates the seizure-enhancing effects of systemic inflammation. J Neurosci 36 (13):3777–3788

27. Kobylarek D, Iwanowski P, Lewandowska Z, Limphaibool N, Szafranek S, Labrzycka A, Kozubski W (2019) Advances in the potential biomarkers of epilepsy. Front Neurol 10:685. https://doi.org/10.3389/fneur.2019.00685

28. Sampath V (2018) Bacterial endotoxin-lipopolysaccharide; structure, function and its role in immunity in vertebrates and invertebrates. Agric Nat Resour 52(2):115–120. https://doi.org/10.1016/j.anres.2018.08.002

29. Abreu MT, Vora P, Faure E, Thomas LS, Arnold ET, Arditi M (2001) Decreased expression of toll-like Receptor-4 and MD-2 correlates with intestinal epithelial cell protection against dysregulated proinflammatory gene expression in response to bacterial lipopolysaccharide. J Immunol 167(3):1609. https://doi.org/10.4049/jimmunol.167.3.1609

30. Layé S, Parnet P, Goujon E, Dantzer R (1994) Peripheral administration of lipopolysaccharide induces the expression of cytokine transcripts in the brain and pituitary of mice. Mol Brain Res 27(1):157–162. https://doi.org/10.1016/0169-328X(94)90197-X

31. Auvin S, Porta N, Nehlig A, Lecointe C, Vallée L, Bordet R (2009) Inflammation in rat pups subjected to short hyperthermic seizures enhances brain long-term excitability. Epilepsy Res 86(2–3):124–130. https://doi.org/10.1016/j.eplepsyres.2009.05.010

32. Sayyah M, Javad-Pour M, Ghazi-Khansari M (2003) The bacterial endotoxin lipopolysaccharide enhances seizure susceptibility in mice: involvement of proinflammatory factors: nitric oxide and prostaglandins. Neuroscience 122:1073–1080. https://doi.org/10.1016/j.neuroscience.2003.08.043

33. Borella TL, De Luca LA Jr, Colombari DSA, Menani JV (2008) Central muscarinic receptor subtypes involved in pilocarpine-induced salivation, hypertension and water intake. Br J Pharmacol 155(8):1256–1263. https://doi.org/10.1038/bjp.2008.355

34. Castro OW, Furtado MA, Tilelli CQ, Fernandes A, Pajolla GP, Garcia-Cairasco N (2011) Comparative neuroanatomical and temporal characterization of FluoroJade-positive neurodegeneration after status epilepticus induced by systemic and intrahippocampal pilocarpine in Wistar rats. Brain Res 1374:43–55. https://doi.org/10.1016/j.brainres.2010.12.012

35. Curia G, Longo D, Biagini G, Jones RSG, Avoli M (2008) The pilocarpine model of temporal lobe epilepsy. J Neurosci Methods 172 (2):143–157. https://doi.org/10.1016/j.jneumeth.2008.04.019

36. Mendes NF, Pansani AP, Carmanhães ERF, Tange P, Meireles JV, Ochikubo M, Chagas JR, da Silva AV, Monteiro de Castro G, Le Sueur-Maluf L (2019) The blood-brain barrier breakdown during acute phase of the pilocarpine model of epilepsy is dynamic and time-dependent. Front Neurol 10:382–382. https://doi.org/10.3389/fneur.2019.00382

37. Scorza FA, Arida RM, Naffah-Mazzacoratti MG, Scerni DA, Calderazzo L, Cavalheiro EA (2009) The pilocarpine model of epilepsy: what have we learned? An Acad Bras Cienc 81:345–365

38. Soond DR, Bjorgo E, Moltu K, Dale VQ, Patton DT, Torgersen KM, Galleway F, Twomey B, Clark J, Gaston JS, Tasken K, Bunyard P, Okkenhaug K (2010) PI3K p110delta regulates T-cell cytokine production during primary and secondary immune responses in mice and humans. Blood 115 (11):2203–2213. https://doi.org/10.1182/blood-2009-07-232330

39. Saha RN, Ghosh A, Palencia CA, Fung YK, Dudek SM, Pahan K (2009) TNF-alpha preconditioning protects neurons via neuron-

39. specific up-regulation of CREB-binding protein. J Immunol (Baltimore, Md: 1950) 183 (3):2068–2078. https://doi.org/10.4049/jimmunol.0801892

40. Marsh BJ, Williams-Karnesky RL, Stenzel-Poore MP (2009) Toll-like receptor signaling in endogenous neuroprotection and stroke. Neuroscience 158(3):1007–1020. https://doi.org/10.1016/j.neuroscience.2008.07.067

41. Mirrione MM, Konomos DK, Gravanis I, Dewey SL, Aguzzi A, Heppner FL, Tsirka SE (2010) Microglial ablation and lipopolysaccharide preconditioning affects pilocarpine-induced seizures in mice. Neurobiol Dis 39 (1):85–97. https://doi.org/10.1016/j.nbd.2010.04.001

42. Akarsu ES, Ozdayi S, Algan E, Ulupinar F (2006) The neuronal excitability time-dependently changes after lipopolysaccharide administration in mice: possible role of cyclooxygenase-2 induction. Epilepsy Res 71 (2–3):181–187. https://doi.org/10.1016/j.eplepsyres.2006.06.009

43. Yuhas Y, Nofech-Mozes Y, Weizman A, Ashkenazi S (2002) Enhancement of pentylenetetrazole-induced seizures by Shigella dysenteriae in LPS-resistant C3H/HeJ mice: role of the host response. Med Microbiol Immunol 190(4):173–178

44. Sankar R, Auvin S, Mazarati A, Shin D (2007) Inflammation contributes to seizure-induced hippocampal injury in the neonatal rat brain. Acta Neurol Scand Suppl 186:16–20

45. Ho Y-H, Lin Y-T, Wu C-WJ, Chao Y-M, Chang AYW, Chan JYH (2015) Peripheral inflammation increases seizure susceptibility via the induction of neuroinflammation and oxidative stress in the hippocampus. J Biomed Sci 22(1):46. https://doi.org/10.1186/s12929-015-0157-8

46. Ko KR, Frampton JP (2016) Developments in 3D neural cell culture models: the future of neurotherapeutics testing? Expert Rev Neurother 16(7):739–741. https://doi.org/10.1586/14737175.2016.1166053

47. Cho T, Bae JH, Choi HB, Kim SS, McLarnon JG, Suh-Kim H, Kim SU, Min CK (2002) Human neural stem cells: electrophysiological properties of voltage-gated ion channels. Neuroreport 13(11):1447

48. Turski WA, Cavalheiro EA, Bortolotto ZA, Mello LM, Schwarz M, Turski L (1984) Seizures produced by pilocarpine in mice: a behavioral, electroencephalographic and morphological analysis. Brain Res 321 (2):237–253. https://doi.org/10.1016/0006-8993(84)90177-X

49. Riss TL, Moravec RA, Niles AL, Duellman S, Benink HA, Worzella TJ, Minor L (2016) Cell viability assays. In: Assay guidance manual [Internet]. Eli Lilly & Company and the National Center for Advancing Translational Sciences

50. O'Brien J, Wilson I, Orton T, Pognan F (2000) Investigation of the Alamar blue (resazurin) fluorescent dye for the assessment of mammalian cell cytotoxicity. Eur J Biochem 267(17):5421–5426

51. Roehm NW, Rodgers GH, Hatfield SM, Glasebrook AL (1991) An improved colorimetric assay for cell proliferation and viability utilizing the tetrazolium salt XTT. J Immunol Methods 142(2):257–265

52. Pacico N, Mingorance-Le Meur A (2014) New in vitro phenotypic assay for epilepsy: fluorescent measurement of synchronized neuronal calcium oscillations. PLoS One 9(1): e84755–e84755. https://doi.org/10.1371/journal.pone.0084755

53. Blümcke I, Spreafico R, Haaker G, Coras R, Kobow K, Bien C, Pfäfflin M, Elger C, Widman G, Schramm J, Becker A, Braun K, Leijten F, Baayen J, Aronica E, Chassoux F, Hamer H, Stefan H, Roessler K, Avanzini G (2017) Histopathological findings in brain tissue obtained during epilepsy surgery. N Engl J Med 377:1648–1656. https://doi.org/10.1056/NEJMoa1703784

54. Mohapel P, Ekdahl CT, Lindvall O (2004) Status epilepticus severity influences the long-term outcome of neurogenesis in the adult dentate gyrus. Neurobiol Dis 15(2):196–205. https://doi.org/10.1016/j.nbd.2003.11.010

55. Nirwan N, Vyas P, Vohora D (2018) Animal models of status epilepticus and temporal lobe epilepsy: a narrative review. Rev Neurosci 29 (7):757–770. https://doi.org/10.1515/revneuro-2017-0086

56. Bengzon J, Kokaia Z, Elmér E, Nanobashvili A, Kokaia M, Lindvall O (1997) Apoptosis and proliferation of dentate gyrus neurons after single and intermittent limbic seizures. Proc Natl Acad Sci U S A 94(19):10432–10437. https://doi.org/10.1073/pnas.94.19.10432

57. Covolan L, Ribeiro LT, Longo BM, Mello LE (2000) Cell damage and neurogenesis in the dentate granule cell layer of adult rats after pilocarpine- or kainate-induced status epilepticus. Hippocampus 10(2):169–180. https://doi.org/10.1002/(sici)1098-1063(2000)10:2<169::aid-hipo6>3.0.co;2-w

58. Ekdahl CT, Mohapel P, Elmér E, Lindvall O (2001) Caspase inhibitors increase short-term survival of progenitor-cell progeny in the adult rat dentate gyrus following status epilepticus.

Eur J Neurosci 14(6):937–945. https://doi.org/10.1046/j.0953-816x.2001.01713.x

59. Nakagawa E, Aimi Y, Yasuhara O, Tooyama I, Shimada M, McGeer PL, Kimura H (2000) Enhancement of progenitor cell division in the dentate gyrus triggered by initial limbic seizures in rat models of epilepsy. Epilepsia 41 (1):10–18. https://doi.org/10.1111/j.1528-1157.2000.tb01498.x

60. Parent JM, Yu TW, Leibowitz RT, Geschwind DH, Sloviter RS, Lowenstein DH (1997) Dentate granule cell neurogenesis is increased by seizures and contributes to aberrant network reorganization in the adult rat hippocampus. J Neurosci 17(10):3727–3738. https://doi.org/10.1523/JNEUROSCI.17-10-03727.1997

61. Watanabe K, Yamaji R, Ohtsuki T (2018) MicroRNA-664a-5p promotes neuronal differentiation of SH-SY5Y cells. Genes Cells: Devoted Mol Cell Mech 23(3):225–233. https://doi.org/10.1111/gtc.12559

62. Eves EM, Tucker MS, Roback JD, Downen M, Rosner MR, Wainer BH (1992) Immortal rat hippocampal cell lines exhibit neuronal and glial lineages and neurotrophin gene expression. Proc Natl Acad Sci U S A 89 (10):4373–4377. https://doi.org/10.1073/pnas.89.10.4373

63. Ross J, Olmsted JB, Rosenbaum JL (1975) The ultrastructure of mouse neuroblastoma cells in tissue culture. Tissue Cell 7 (1):107–135. https://doi.org/10.1016/S0040-8166(75)80010-3

64. Kovalevich J, Langford D (2013) Considerations for the use of SH-SY5Y neuroblastoma cells in neurobiology. Methods Mol Biol (Clifton, NJ) 1078:9–21. https://doi.org/10.1007/978-1-62703-640-5_2

65. Weksler B, Romero IA, Couraud PO (2013) The hCMEC/D3 cell line as a model of the human blood brain barrier. Fluids Barriers CNS 10(1):16. https://doi.org/10.1186/2045-8118-10-16

66. Grainger AI, King MC, Nagel DA, Parri HR, Coleman MD, Hill EJ (2018) in vitro models for seizure-liability testing using induced pluripotent stem cells. Front Neurosci 12:590. https://doi.org/10.3389/fnins.2018.00590

67. Paavilainen T, Pelkonen A, Mäkinen MEL, Peltola M, Huhtala H, Fayuk D, Narkilahti S (2018) Effect of prolonged differentiation on functional maturation of human pluripotent stem cell-derived neuronal cultures. Stem Cell Res 27:151–161

68. Kayama T, Suzuki I, Odawara A, Sasaki T, Ikegaya Y (2018) Temporally coordinated spiking activity of human induced pluripotent stem cell-derived neurons co-cultured with astrocytes. Biochem Biophys Res Commun 495 (1):1028–1033

69. Ishii MN, Yamamoto K, Shoji M, Asami A, Kawamata Y (2017) Human induced pluripotent stem cell (hiPSC)-derived neurons respond to convulsant drugs when co-cultured with hiPSC-derived astrocytes. Toxicology 389:130–138

70. Auvin S, Shin D, Mazarati A, Sankar R (2010) Inflammation induced by LPS enhances epileptogenesis in immature rat and may be partially reversed by IL1RA. Epilepsia 51(Suppl 3):34–38. https://doi.org/10.1111/j.1528-1167.2010.02606.x

71. Eun BL, Abraham J, Mlsna L, Kim M, Koh S (2015) Lipopolysaccharide potentiates hyperthermia-induced seizures. Brain Behav 5:e00348. https://doi.org/10.1002/brb3.348

72. Verdaguer E, García-Jordà E, Jiménez A, Stranges A, Sureda FX, Canudas AM, Escubedo E, Camarasa J, Pallàs M, Camins A (2002) Kainic acid-induced neuronal cell death in cerebellar granule cells is not prevented by caspase inhibitors. Br J Pharmacol 135 (5):1297–1307. https://doi.org/10.1038/sj.bjp.0704581

73. Tsai H-L, Chang S-J (2015) Key proteins of activating cell death can be predicted through a kainic acid-induced excitotoxic stress. Biomed Res Int 2015:478975. https://doi.org/10.1155/2015/478975

74. Bibi F, Ullah I, Kim MO, Naseer MI (2017) Metformin attenuate PTZ-induced apoptotic neurodegeneration in human cortical neuronal cells. Pak J Med Sci 33(3):581–585. https://doi.org/10.12669/pjms.333.11996

75. Ta T-T, Dikmen HO, Schilling S, Chausse B, Lewen A, Hollnagel J-O, Kann O (2019) Priming of microglia with IFN-γ slows neuronal gamma oscillations in situ. Proc Natl Acad Sci 116(10):4637. https://doi.org/10.1073/pnas.1813562116

76. Park KM, Yule DI, Bowers WJ (2008) Tumor necrosis factor-alpha potentiates intraneuronal Ca2+ signaling via regulation of the inositol 1,4,5-trisphosphate receptor. J Biol Chem 283(48):33069–33079. https://doi.org/10.1074/jbc.M802209200

77. Kruger NJ (2009) The Bradford method for protein quantitation. In: The protein protocols handbook. Springer, Totowa, New Jersey, pp 17–24

78. Steiner AA, Molchanova AY, Dogan MD, Patel S, Pétervári E, Balaskó M, Wanner SP,

Eales J, Oliveira DL, Gavva NR, Almeida MC, Székely M, Romanovsky AA (2011) The hypothermic response to bacterial lipopolysaccharide critically depends on brain CB1, but not CB2 or TRPV1, receptors. J Physiol 589 (Pt 9):2415–2431. https://doi.org/10.1113/jphysiol.2010.202465

79. Bailey SA, Zidell RH, Perry RW (2004) Relationships between organ weight and body/brain weight in the rat: what is the best analytical endpoint? Toxicol Pathol 32(4):448–466. https://doi.org/10.1080/01926230490465874

Chapter 4

Cellular Electrophysiological Methods to Decipher the Altered Synaptic Transmission Associated with Drug-Resistant Epilepsy

Soumil Dey, Aparna Banerjee Dixit, Manjari Tripathi, P. Sarat Chandra, and Jyotirmoy Banerjee

Abstract

Drug-resistant epilepsy (DRE) affects almost one-third of the patients having epilepsy. These patients suffer from unprovoked seizures which are refractory to antiepileptic drugs, and therefore surgical intervention is the only therapeutic option. The root cause of seizure generation is not perfectly understood, but the prevailing theory is imbalance between excitatory and inhibitory neurotransmission within the brain. In this chapter, we have described cellular electrophysiological approaches to investigate abnormal synaptic transmission in neurons of the surgically resected brain tissues obtained from the DRE patients. This includes functional activity of the glutamatergic and GABAergic receptors in resected brain tissues obtained from patients with hippocampal sclerosis, the most common DRE pathology using patch clamp electrophysiology. Our aim is to provide a guideline on how to apply the cellular electrophysiological methods to study altered synaptic transmission associated with DRE and its application to identify drug-resistant epileptogenic network.

Key words Drug-resistant epilepsy, Patch clamp technique, Synaptic transmission, Postsynaptic current, Hippocampal sclerosis

1 Introduction

Epilepsy is a chronic neural disease with primary indication of recurrent unprovoked seizures essentially due to the excessive discharge from neurons [1]. According to WHO, epilepsy accounts for almost 1% of the total global disease burden [2]. Regardless of the various types of anti-epileptic drugs (AEDs) used, almost 30% of all the epilepsy patients do not achieve complete seizure freedom with the AEDs alone. According to the International League Against Epilepsy (ILAE), drug-resistant epilepsy (DRE) is when two adequately selected, tolerated, and used anti-epileptic drugs either individually or in combination fail to attain complete

Divya Vohora (ed.), *Experimental and Translational Methods to Screen Drugs Effective Against Seizures and Epilepsy*, Neuromethods, vol. 167, https://doi.org/10.1007/978-1-0716-1254-5_4,
© Springer Science+Business Media, LLC, part of Springer Nature 2021

freedom from recurrent seizure for a duration of not more than 2 years in adults and in pediatrics much earlier (within weeks of onset of seizures) [3]. If the patient remains free of seizure for more than a year or suffers from occasional seizures separated by a duration three times the longest interval between two seizure episodes before the proper treatment or intervention, then complete freedom from seizure is considered to be achieved [4]. Although surgical intervention is recommended for the treatment of DRE, due to intricate origin and complex pathologies, surgical intervention does not always achieve a comprehensive outcome [5].

To study abnormal synaptic transmission associated with DRE, patch clamp recordings from the neurons of the brain slices obtained from the patients with epilepsy are the most widely used technique that permits the electrophysiological measurements of ion channel currents present in the cell membrane. Since its inception in 1976 by Erwin Neher and Bert Sakmann, this is the most preferred and useful technique for the measurements of ion-channel activities in cells with resolution of up to single type of ion channel [6]. This technique has provided important understanding into the origin of seizures and hyperexcitability associated with drug-resistant epilepsy, and failure of action of anti-epileptic drugs.

The genesis of DRE is still not completely understood, but a key element is altered synaptic transmission. Generation of seizure is primarily due to development of a hyperexcitable neuronal network because of disruption of balance between excitatory and inhibitory neurotransmission. To this end, cellular electrophysiological studies of surgically resected epileptic brain tissue obtained from DRE patients for the intracellular electrophysiological analysis help to investigate abnormal synaptic transmission. In addition, they also help to determine the contribution of a specific neurotransmitter-mediated synaptic transmission through pharmacological intervention.

The traditional simplified concept of "epileptic zone" where a single and static epileptic lesion was considered to be surrounded by anatomically normal but physiologically deranged cortical tissues has rather changed into the concept of multiple and dynamic "epileptogenic network" which increase in size and complexity over time. A cohort of neurons with bi-stable firing pattern epileptogenically stimulate some distant group of neurons which reciprocate and form a reverberating circuit also called "node." These nodes gradually recruit more and more distant neuronal groups and form larger circuits, and the epileptogenic network expands over time [7–10]. So, concurring to this concept of epileptogenic network, an epileptogenic focus can be present distantly from epileptic lesion separated by normal brain parenchyma. Neural network hypothesis for DRE postulates that recurrent abnormal neuronal hyperactivity contributes in generating abnormal synaptic reorganization, abnormal connection with inferior synapses, axonal sprouting, gliosis, and neuronal network remodeling. Gradually, novel pathological networks are formed, and epileptogenic focus changes its course toward new non-physiological direction. This new neural network becomes not only resistant to endogenous anti-epileptic system but also ineffective to external anti-epileptic drugs, and consequently forms DRE [11]. The abnormal neural network is well documented in case of mesial temporal lobe epilepsy with hippocampal sclerosis (HS) where the dentate gyrus region of the hippocampus undergoes substantial anatomical reorganization. The most prominent feature of this is the formation of recurrent excitatory collaterals between dentate gyrus granule cells via mossy-fiber sprouting [12, 13]. This reverberating circuit is sufficient to create hyperexcitability in response to normal perforant pathway input [14].

Modern medicine and translational science now have the great advantage of having reproducible animal model for human disease which is analogous and equivalent to human disorder in every aspect. Similarly, over the last few decades many animal models of epilepsy (mostly temporal lobe epilepsy model) have been developed which replicate behavioral, electroencephalographic, and neuropathological characteristics of human DREs. DRE is a distributed neuronal network disorder, and it is possible that the multiple epileptogenic networks located in different regions of the brain are responsible for seizure generation. Thus, it is warranted that synaptic transmission is measured in the samples obtained from different regions of the brain. Since, due to ethical constraints it is not possible to receive region-specific specimens from patients with DRE, animal models of epilepsy give us the opportunity to understand the alteration in synaptic transmission from samples obtained from different brain regions. To fully understand the intricate pathology of epileptogenesis, validation of the experimental findings from the DRE patients as well as for intervention of new therapeutic agents, many animal models have been created in which initial brain lesions are induced either physically or by chemiconvulsants. In this chapter, we have demonstrated electrophysiological studies on samples obtained from patients with HS and the pilocarpine rodent model of temporal lobe epilepsy.

2 Materials

2.1 Patients and Pre-surgical Evaluations

The experiments discussed here are approved by the Institutional Ethics Committee, All India Institute of Medical Sciences (AIIMS), New Delhi. All the patients with DRE undergo several pre-surgical standard workups which include 3 T epilepsy protocol MRI, and concordant data from video EEG (vEEG), fluoro-2-deoxyglucose positron emission tomography (FDG-PET), single-photon emission computed tomography (SPECT), magnetoencephalography (MEG) evaluations, and intraoperative electrocorticography (ECoG) [15–18]. Prior to surgery, details of each patient are discussed by epilepsy surgeons, epileptologists, neuro-radiologists, and nuclear medicine specialists, and competence of the epilepsy patients for surgery is determined based on the converging data of the above-mentioned multimodal imaging techniques. Postoperative histopathological examinations are performed for confirmation of underlying pathologies in consonance with the recent classification provided by the World Health Organization.

Ideally, the non-seizure controls for DRE patients will be representative areas of the brain from healthy individuals, but this is not possible due to ethical reasons. Conceptually, human brain tissues of similar ages resected from non-seizure pathology such as low-grade gliomas or samples from trauma patients serve as potential control tissue, but however it is not viable to get region-matched tissues from the ethical perspective. Unfortunately, the tissues isolated from the trauma patients are usually severely damaged or contused, so they do not serve the purpose. Thus, in our study we have used resected brain specimens obtained from peripheral areas of tumors during surgical resection which is also a part of planned surgical resection in low-grade glioma patients without any seizure histories as non-seizure controls [19–21].

2.2 Patch Clamp Rig

Due to the wide variation in experimental setups and categories of experiment that can be designed, different arrays of patch clamp setups are used in various laboratories. A standard patch clamp rig consists of a patch clamp amplifier along with a data recording software, a micromanipulator holding the amplifier headstage containing a patch pipette for precise positioning on to the cell, and a phase contrast microscope (for live neuron localization in a slice preparation) kept on a vibration-free table, and the whole system is covered by a Faraday cage. The rig that is used for most of the studies mentioned in this chapter is comprised of a feedback amplifier (Axopatch 200B, Molecular Devices, USA) and a CV203BU headstage. The amplifier is connected to a digitizer (Digidata 1440A, Molecular Devices) for converting analogue data into digital one and which in turn is recorded into a computer hard disk using the software PCLAMP 10.0 (Molecular Devices). To visualize the live neurons within slices, we have used an upright microscope (fixed stage) which is equipped with differential interference contrast (DIC) optics. For visualization, we have used either a CFI Plan Fluor $10\times/0.30$ N.A. objective or a CFI Plan Apochromat NIR $40\times/0.80$ N.A. water immersion objective (Eclipse FN1, Nikon, Japan). The light source is a pre-centered 12 V–100 W long life halogen lamp. A CCD camera (C3077–80, Hamamatsu, Japan) is attached to the microscope for video recording of the neuronal imaging. In the near-infrared region, this camera is highly sensitive, and in comparison to the conventional model, there is more than twofold increase in spectral response at 900 nm. The fixed stage (MY-1078, Sutter Instruments, USA) is attached with the microscope equipped with an insert designed to accommodate 35 mm petri dishes which acts as a recording chamber where a brain slice is placed and kept stable using a slice anchor. The recording chamber is constantly perfused with artificial cerebrospinal fluid. The microscope is placed on an X-Y translator which allows it to move with respect to the fixed stage. The headstage is connected to a micromanipulator (PatchStar, Scientifica, UK) placed on the fixed stage for accurate placement of recording and stimulating electrodes. The recording chamber is also connected to a drug delivery system (VC-6 six channel perfusion valve control system, Harvard Apparatus, USA). The whole setup is placed on a vibration isolation Table (1200 × 900 mm, TMC, USA) and covered by a Faraday cage which are used to eliminate noise caused by mechanical vibration and electromagnetic fields, respectively. To prevent any residual electrical noises, the headstage and its output wire up to the amplifier is wrapped with aluminum foil. To shield the sensitive experiments from electrical noise, all metallic facades (the vibration isolation table, micromanipulators, microscope, stage, Faraday cage, etc.) are connected to a floating ground, using low-resistance copper ground cables, which is connected to earthing ground (*see* **Note 4.3**).

The glass capillary tubes that are used for preparing micropipettes are 100 mm in length, 0.68 mm diameter on the inner side, 1.2 mm diameter on the outer side, and fire polished (Model 1B120-F4, World Precision Instruments, Florida, USA). P-97 Flaming/Brown type horizontal micropipette puller (Sutter Instruments, USA) is used to fabricate micropipettes from the glass capillary tubes. Each micropipette is filled with internal pipette solution using a copper filler (Microfil, WPI) through 0.22μm syringe filter. The other end of the micropipette is connected to an Ag/AgCl electrode (a thin silver wire electrode was chloride coated all through its entire length) and finally to the amplifier headstage. A bare chloride-coated silver wire reference electrode is placed into the bath (extracellular space) within the petri dish immersed into the solution, and the difference in electrical potential between the two electrodes is measured.

2.3 Solutions and Reagents

The brain specimens are stored in artificial cerebrospinal fluid (ACSF), constantly bubbled with carbogen (95% O_2 + 5% CO_2), immediately after resection from the patients undergoing epilepsy surgery. This solution maintains osmolality, pH, and source of energy for the neurons. The ACSF consists of the following composition (*see* **Note 4.6**): sodium chloride (NaCl), 125 mM; sodium bicarbonate ($NaHCO_3$), 25 mM; potassium chloride (KCl), 2.5 mM; sodium dihydrogen orthophosphate (NaH_2PO_4),

1.25 mM; calcium chloride ($CaCl_2$), 2 mM, magnesium chloride ($MgCl_2$), 1 mM; and glucose ($C_6H_{12}O_6$), 25 mM. The internal pipette solution consists of the following constituents: 4-(2-hydro-xyethyl)-1-piperazineethanesulfonic acid, 10 mM; ethylene-glycol bis (β-amino-ethyl ether)-N-N'-tetra acetic acid, 10 mM; CsCl, 10 mM; Cs-methane sulfonate, 130 mM; $MgCl_2$, 2 mM (CsOH is used to adjust pH to 7.3–7.4; *see* **Note 4.7**). Cs-methane sulfonate pushes chloride reversal potential to more positive value so that spontaneous glutamate receptor currents can be recorded at −70 mV without any interference of GABA receptors.

As NMDA and AMPA receptor antagonists, we have used 2-amino-5-phosphonovaleric acid (APV) and 6-cyano-7-nitroqui-noxalene-2,3-dione (CNQX), respectively. (−) Bicuculline methchloride is used for its antagonistic action on $GABA_A$ receptor, and for suppressing the action potential, tetrodotoxin citrate (TTX) is used.

To develop an animal model of temporal lobe epilepsy, pilocarpine, a muscarinic agonist, is used to induce status epilepticus; lithium chloride to sensitize the brain for pilocarpine action; and methyl scopolamine, a muscarinic antagonist, to inhibit peripheral parasympathetic effects due to pilocarpine [22]. The details are also available in Chapter 7. All of the abovementioned reagents are procured from Sigma Aldrich (MO, USA) except TTX is purchased from Alomone lab (Israel).

2.4 Animal Model of Temporal Lobe Epilepsy

In this model, systemic administration of pilocarpine initiates status epilepticus, followed by a latent period and then spontaneous recurrent seizures, which typically reproduces the imbalance in synaptic transmission associated with human DREs. The experiments discussed here are approved by the Institutional Animal Ethics Committee, All India Institute of Medical Sciences (AIIMS), New Delhi. Male Sprague-Dawley rats (weight 200–250 g) are used to develop the acute Li-pilocarpine model of temporal lobe epilepsy (TLE) [22]. For this purpose, 24 h prior to pilocarpine injection, intraperitoneally lithium chloride (LiCl 127 mg/kg body weight; BW) is administered to increase the sensitivity of the rats to pilocarpine which is due to lithium-induced increase in peripheral inflammatory cytokines such as interleukin-1β and concomitant blood–brain barrier disruption. On Day 2, initially methyl scopolamine (1 mg/kg BW) is injected intraperitoneally to reduce peripheral parasympathetic actions, and after 30 min pilocarpine (240 mg/kg body weight) is administered. The rats are monitored and the seizures were classified into five stages according to Racine's scale: (1) movement of mouth and face, (2) nodding of the head, (3) clonus of the forelimbs, (4) clonus and rearing of the forelimbs, (5) clonus, rearing of the forelimbs and falling. When the rats arrive at stage 4 and 5 for more than 30 min, then it is considered as status epilepticus (SE) [23]. SE induction is interrupted after 90 min of its initiation by intraperitoneal administration of anticonvulsant diazepam (10 mg/kg, body weight) to rescue the animal [24, 25].

3 Methods

3.1 Patient Sample Collection and Experimental Procedure

Based on the concordant data from all the pre-surgical evaluations and intraoperative ECoG recording, tailored surgical procedures are performed to remove the epileptic foci from the brains of the DRE patients. This resected brain tissue is collected from the operation room and immediately placed into a beaker containing well-carbogenated ACSF and brought to the laboratory (*see* **Note 4.5**). ACSF needs to be freshly prepared before every experiment. The resected brain specimens obtained from patients with DRE is resized with a sterile blade, and all gliotic portions and blood vessels are removed and fixed on an agar block which is prepared by dissolving 3% agar (Bactoagar, BD Biosciences) in distilled water. These blocks are used for preparing 350–400µm thick slices using vibrating blade microtome (VT1000S, Leica, Germany). Temperature of the vibratome bath is kept at 4 °C using a recirculating chiller (Julabo, UK). The tissue slices are incubated at room temperature for 30 mins prior to the recordings (Fig. 1a, b; *see* **Note 4.1**).

Cortical slice preparations are placed in the recording chamber using a slice anchor for experiment where it is continuously perfused with well-carbogenated ACSF at a rate of 2 mL/min. The 40× water dipped objective of the microscope is used to visually identify pyramidal neurons which have pyramid-shaped cell body or

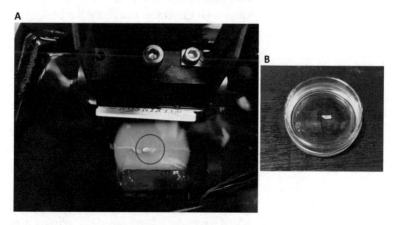

Fig. 1 Slice preparation from resected cortical tissues obtained from patients with DRE undergoing surgery at operation room: (**a**) the epileptic cortical tissue was placed in between two pieces of agar blocks which were adhered to the plexiglass holder for slicing, and 350µM thick slices were prepared in the vibratome bath filled with ice-cold carbogenated ACSF. (**b**) A single cortical slice placed inside ACSF prior to recording

soma and only one prominent tapering apical dendrite. A freshly prepared glass micropipette is filled with internal pipette solution (*see* **Note 4.4**). It is placed within the electrode holder containing Ag/AgCl electrode and attached to the headstage of the amplifier (*see* **Note 4.2**). The micromanipulator wheel is used to move the micropipette toward the cell surface. A brief positive pressure is applied to the micropipette while moving to avoid clogging of the micropipette tip. When pipette tip is inside the bath fluid, a low-voltage (5 mV) square pulse is applied from the feedback amplifier. When filled with internal pipette solution, the glass micropipette inside the bath shows resistances between 4 and 6 MΩ. The amplitudes of current and resistance of this pulse are monitored to follow the formation of the seal. Simultaneously, pipette offset potential (offset potential between reference electrode and micropipette) is neutralized to 0 mV from the amplifier. When the pipette tip is positioned on the cell surface, there is an increase in the resistance which produces, in accordance with Ohm's law ($I = V/R$ where V = voltage, I = current, and R = resistance), a decrease in the current. Then the positive pressure is withdrawn which results in the formation of a giga ohm ($10^9\ \Omega$) seal between the cell membrane and the pipette. The tight giga ohm seal mechanically stabilizes the connection between the micropipette tip and the membrane and amenable to manipulations that allow the establishment of different recording configurations. The whole-cell patch clamp configuration is achieved by applying suction through micropipette holder.

For the animal model of temporal lobe epilepsy, following rescue from SE, the rats are euthanized using carbon dioxide and decapitated using a guillotine. The brains are isolated immediately and placed into the well-carbogenated vibratome bath. The hippocampus is dissected out from the brain, and 350–400μm thick slices are prepared. One slice is placed inside the recording chamber, and a single neuron is patched in whole-cell configuration in the same way as mentioned above.

Step-by-Step Procedure/Methodology:

1. Brain tissues from DRE patients are collected from operation room and immediately brought to laboratory.

2. Following induction of lithium-pilocarpine model of temporal lobe epilepsy, adult male Sprague-Dawley rats are sacrificed, the entire brain is isolated, and the hippocampus is dissected out.

3. Tissues are kept inside a vibratome bath chamber (ice-cold ACSF and well carbogenated).

4. 350–400μm thick slices are prepared using vibratome blade.

5. Slices are incubated for around 30 min at room temperature in well-carbogenated ACSF.

6. One slice is placed inside patch clamp recording chamber and held using a slice anchor.

7. The slices should be continuously perfused with ACSF at a flow rate of 2 mL/min.

8. Pyramidal neurons within the slice are located using IR-DIC microscopy.

9. A glass micropipette is filled with internal pipette solution, attached to the amplifier headstage, and moved toward an already selected pyramidal neuron using micromanipulator controller.

10. A little positive pressure is applied into the glass micropipette to avoid any clogging of the micropipette tip. This also helps in the formation of the giga ohm seal.

11. A 5 mV pulse protocol is initiated through the amplifier during positioning of the micropipette toward the pyramidal neuron.

12. Any pipette offset is corrected with the help of amplifier.

13. When the micropipette tip touches with the cell surface, the resistance increases, and consequently the current (height of the pulse protocol) decreases as per Ohm's law.

14. The positive pressure from micropipette tip is released which creates a tight giga ohm resistance seal between the cell membrane and micropipette tip.

15. Further suction is applied to rupture the cell membrane and achieve whole-cell configuration.

16. The neuron is voltage clamped at -70 mV to record spontaneous postsynaptic currents from glutamate receptors and clamped at 0 mV to record spontaneous postsynaptic currents from GABA receptors.

17. Suitable blockers are used during recording from a particular type of channel (APV and CNQX for recording GABA receptors and bicuculline for glutamate receptors).

18. All experiments are performed at room temperature (20–22 °C).

3.2 Data Acquisition and Interpretation

After whole-cell patch clamp configuration is achieved, voltage clamp mode is used to record spontaneous postsynaptic currents. The neurons are clamped at -70 mV for recording of excitatory postsynaptic currents (EPSCs) and at 0 mV for those inhibitory postsynaptic currents (IPSCs). During recording of EPSCs, 10μM bicuculline is added into the ACSF/extracellular fluid to inhibit $GABA_A$ receptors, and while recording of IPSCs, 50μM APV and 10μM CNQX are added into the ACSF for blocking NMDA and

AMPA/kainite type glutamate receptors. Data are acquired in Clamplex module of the pCLAMP 10.0 software for a 5 min epoch. Signals are filtered at 1 kHz and digitized through Digidata 1440A at 10 kHz. The leak current should be ideally within 50–150 pA, but when it crosses 200 pA, the data are not included for the analysis (*see* **Note 4.9**). The access resistance is monitored during the entire course of the experiment and ranges between 15 and 20 MΩ (*see* **Note 4.8**). Data from any of the neuron are not considered for analysis when the access resistance increases more than 20% from the initial value. Data are analyzed post recording using Clampfit module of the same software. After opening the raw data file into Clampfit analysis window, several kinds of data conditioning, like filtering and baseline correction, need to be performed. Filtering includes notch and electrical interference filters in addition to conventional high and low pass options. Different parameters of synaptic events, namely, frequency, peak amplitude, rise time (10–90%), and decay-time constant (τ_d), are also measured. For this purpose, amplitude baseline and threshold levels are set, which for EPSCs are at −5 pA and for IPSCs are at +10 pA, and then the epochis searched for data that cross the thresholds. The events which cross the threshold value are counted manually. Each event should contain a steep rising phase and an exponential decay phase (Fig. 2e). Events that do not show the above-mentioned rise and decay phase are not typical synaptic waveform, and these are rejected manually (*see* **Note 4.10**). Double- and multiple-peak events are included for calculation of frequency of EPSCs as multiple events, but are excluded for the determination of kinetic properties of the EPSCs/IPSCs. For analysis of kinetic properties which are rise time (10–90%) and decay time constant, only single events which show a steep or sharp rising phase and an exponential decay are chosen during manual analysis of the recordings. The rise time is a current step function, i.e., it is the time required for EPSCs to go from certain low value to a certain high value. In other words, it's the opening kinetics time of ion channels during generation of EPSCs or IPSCs. It is defined as the time required for the synaptic events to rise from 10% to 90% of its peak amplitude. Similarly decay time constant or τ_d shows the closing kinetics of ion channels. As behaviors of neurons are similar to that of R-C circuits, the τ_d is the time required to discharge the capacitor (lipid bilayer) through the resistor (ion channels) to 36.8% of its peak amplitude current. The τ_d are calculated for synaptic events that clearly show an exponential decay phase and fit to exponential phase equation (Fig. 2e). After calculating all the synaptic events manually within every 5 min epoch, all the measurements of all selected synaptic events are averaged and used as final values for the total epoch. Furthermore, the cumulative distributions of inter-event intervals and peak amplitude between groups are compared by using the Kolmogorov-Smirnov (K-S) test. Events

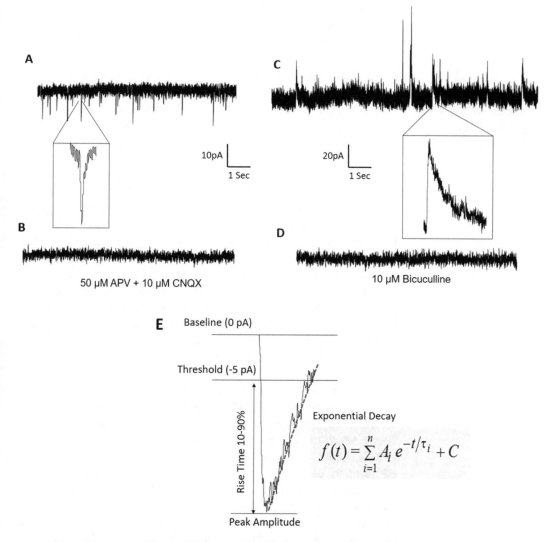

Fig. 2 Recordings of spontaneous EPSCs and IPSCs from pyramidal neurons in non-seizure controls. (**a**) Sample recordings of spontaneous EPSCs recorded from pyramidal neurons in cortical sample obtained from non-seizure controls. Inset shows a single EPSC event at an expanded time scale. (**b**) The second trace shows absence of any spontaneous EPSCs following perfusion of the slice with glutamate receptor antagonists APV (50μM) and CNQX (10μM) for 10 min proving that these events are glutamate receptor mediated. (**c**) Sample recordings of spontaneous IPSCs recorded from pyramidal neurons in cortical sample obtained from the non-seizure controls. Inset shows a single IPSC event at an expanded time scale. (**d**) The fourth trace shows absence of any spontaneous IPSCs following perfusion of the slice with GABA$_A$ receptor antagonists 10μM bicuculline for 10 min proving that these events are mediated by GABA$_A$ receptors. (**e**) Sample analysis of an EPSC event showing a sharp rise and an exponential decay phase. After adjusting the baseline, a threshold is drawn which is essentially at −5 pA. The time to reach from the baseline to the peak amplitude is the rise time. The decay phase should be exponential, for which this phase needs to be fitted with a standard exponential equation which is shown along with figure

from different neurons in each group are pooled together and then subjected to the K-S test by using the Clampfit module of pCLAMP software.

3.3 Recording of Glutamatergic Synaptic Activity from Samples Obtained from Patients with Hippocampal Sclerosis

Hippocampal sclerosis is the most common form of DRE where unprovoked seizure generation occurs from extensive epileptogenic network that originates from hippocampus as well as extra-hippocampal areas [26]. Stereo electroencephalography (SEEG) studies of HS patients revealed that sites of seizure onset and propagation during ictal period might be present in two or more mesial or lateral temporal, and/or even in extra-temporal, structures in addition to the hippocampus [27]. MRI and ictal SEEG recordings from patients with HS have indicated two epileptogenic networks that are present in both the hippocampal and extra-hippocampal areas of the temporal cortex [28]. Functional MRI (fMRI) investigations have revealed that even the impromptu resting state brain network activity involved in higher order brain functions are also affected in temporal lobe epilepsy [29]. Recently resting-state MEG analysis also showed network hubs in both the hippocampal and the temporal cortex areas in HS patients reflecting pathophysiologic brain network reorganization [30].

Glutamate and GABA are the chief excitatory and inhibitory neurotransmitters, respectively. A pyramidal neuron of the cerebral cortex, in response to an excitatory input, produces an excitatory postsynaptic potential (EPSP) mediated by glutamate followed by an inhibitory postsynaptic potential (IPSP) mediated by GABA. Every neuron inside the seizure focus has a sudden, long-lasting (50–200 ms), and large (20–40 mV) depolarization plateau in addition to an array of overshooting action potentials, known as paroxysmal depolarizing shift (PDS), synchronized responses of which ultimately result in epileptiform discharges [31]. As long as the abnormal electrical activity is restricted to a small group of neurons, there are no clinical manifestations. The synchronization of the neurons in the focus is dependent not only on the intrinsic properties of each individual cell but also on the connections between those neurons. During the period between two epileptic seizures (interictal period), the PDS is localized to the seizure focus with an inhibitory surrounding which is particularly dependent on the feedback and feed-forward inhibition by the GABAergic inhibitory interneurons [32, 33]. During the focal seizure (ictal period) the inhibitory surround is overcome, and as a result, the continuous high-frequency train of action potentials is generated, and seizure begins to spread beyond the original focus.

Meta-analysis suggests that surgical procedure for the MTLE-HS patients involving resection of the anterior temporal lobe (ATL), combined with amygdalohippocampectomy, have a better outcome compared to standard amygdalohippocampectomy [34]. The standard temporal lobectomy involved removal of the hippocampus and the ATL. We performed whole cell patch clamp recordings from normal looking pyramidal neurons of both ATL and the hippocampus.

We have recorded spontaneous glutamatergic EPSCs from pyramidal neurons in non-seizure controls and (Fig. 2a) hippocampal slice preparations at -70 mV holding potential under whole-cell voltage clamps configuration. The characteristics of the typical EPSCs are sharp rise followed by exponential decay phase (Fig. 2a inset). These events are completely abolished after perfusion of the tissue slices with the ACSF containing the mixture of the NMDA receptor blocker APV (50μM) and the AMPA receptor blocker CNQX (10μM) for 10 min (Fig. 2b). These results suggest that spontaneous EPSCs are mediated by the NMDA as well as AMPA/kainite type glutamate receptors. We observe that the frequencies as well as the amplitudes of the spontaneous EPSCs recorded from the pyramidal neurons of the tissue slices received from the hippocampal sclerosis patients are significantly higher in comparison to that of non-seizure control patients, suggesting hyperexcitation. The frequencies as well as the amplitudes of these spontaneous EPSCs remain unaltered by perfusion of the tissue slices with $GABA_A$ receptor blocker bicuculline (10μM). However, the kinetic properties are not significantly altered between the two groups (Table 1). This suggests that the glutamate receptor-mediated spontaneous activities obtained from the pyramidal neurons are not under influence of the basal GABAergic response at -70 mV. Now, to assess the contribution of the endogenous NMDA receptors in generating hyperexcitation, the tissue slices are perfused with 50μM APV dissolved in ACSF. The frequencies of the spontaneous EPSCs are significantly reduced in both non-seizure control and HS groups; but the percentage reduction of the frequencies is significantly higher in HS in comparison to non-seizure control. In HS patients, at resting condition, enhanced inhibition by APV implies that the NMDA receptors are spontaneously hyperactive at -70 mV, and this enhanced NMDA receptor-mediated glutamatergic tone contributes significantly to the generation of the hyperexcitation. Then

Table 1
Kinetic parameters of spontaneous EPSCs and IPSCs recorded from pyramidal neurons in non-seizure control tissues

Parameters	EPSCs	IPSCs
Frequency (Hz)	0.68 ± 0.04	1.6 ± 0.3
Amplitude (pA)	12.43 ± 1.9	21.3 ± 2.5
Rise time (ms)	1.9 ± 0.7	2.1 ± 0.5
Decay time constant (ms)	9.6 ± 1.4	32.4 ± 3.8

we record spontaneous GABAergic IPSCs at 0 mV holding potential. Unlike EPSCs, spontaneous IPSCs are long duration outward currents with a sharp peak and a long exponential decay phase (Fig. 2c). These events are completely blocked following perfusion of 10μM bicuculline which is a $GABA_A$ receptor blocker (Fig. 2d). But we do not observe much variation in frequency and amplitude of these events between HS hippocampus and non-seizure controls. This suggests that the imbalance between excitation and inhibition observed in HS is primarily due to the enhanced excitatory glutamate receptor-mediated activities rather than GABA receptor-mediated inhibitory activities [20].

We then compare the regional variation of spontaneous EPSCs in the HS patients. Spontaneous EPSCs are recorded at −70 mV under whole-cell voltage clamp mode from the pyramidal neurons of the tissue slices received from the resected sclerotic hippocampus as well as ATL of these patients (Fig. 3a). These currents are completely abolished after perfusion with ACSF containing the mixture of APV (50μM) and CNQX (10μM) for 10 min. The frequencies and amplitudes of the spontaneous EPSCs recorded from the pyramidal neurons in the hippocampus and ATL tissues received from the HS patients are significantly higher in comparison to that of the non-seizure controls (Fig. 3b). In addition to this, the frequencies and amplitudes in ATL tissues are also significantly higher in comparison to the hippocampus. However, the kinetic properties in HS patients are not significantly different from those in non-seizure controls (Tables 2, 3 and 4). Next, these tissues are perfused with the ACSF containing action potential blocker TTX (200 nM) for 10 min. TTX blocks all the action potential-dependent components of EPSCs, which results in reduction of the frequency as well as the amplitude of the spontaneous EPSCs and only the miniature EPSCs remain (Fig. 3c). The frequency and amplitude of the miniatures are comparable between the ATL, hippocampus, and non-seizure control tissues (Fig. 3d). However, the percentage reduction of the frequency and amplitude is more in ATL compared to hippocampi after TTX bath perfusion which implies that contribution of the action potential-dependent component of spontaneous EPSCs is higher in ATL compared to hippocampus. The kinetic properties are not significantly altered by TTX in the hippocampal and ATL samples as compared to the non-seizure controls (Tables 2, 3, and 4). Finally, the GABAergic IPSCs do not vary much between ATL, hippocampus, and non-seizure controls. As described previously [19], two corollaries can be deduced from these findings. First, in HS patients two separate resting state networks are present in ATL and hippocampus, and second, two independent cellular mechanisms differentially mediate spontaneous glutamatergic tone in these two regions. As the GABA receptor activities do not alter much between HS ATL, hippocampus, and non-seizure control, hyperactive

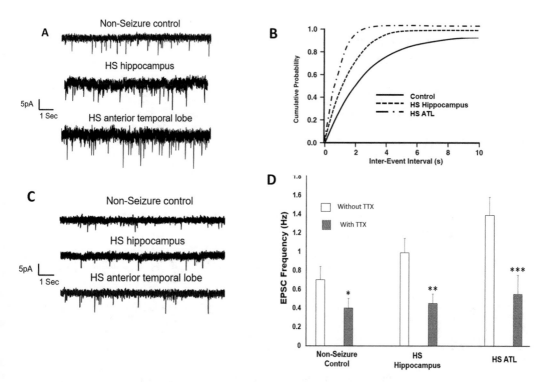

Fig. 3 Spontaneous EPSCs recorded from pyramidal neurons in resected brain tissues obtained from non-seizure controls and HS patients' hippocampus and ATL. (**a**) Sample recording of spontaneous EPSCs recorded from pyramidal neurons of cortical tissues obtained from non-seizure controls, HS hippocampus, and ATL tissues. (**b**) The cumulative probability plot of inter-event interval (seconds) of EPSCs recorded from brain tissues of non-epileptic control, HS hippocampus, and ATL. Plot shows data obtained from seven neurons from six non-seizure control patients, six neurons from six ATL, and ten neurons from nine hippocampus of HS patients. Compared to non-seizure controls in HS hippocampus and ATL, significant leftward displacements of cumulative distribution of inter-event interval ($p < 0.001$) were observed. Statistical analysis for cumulative distribution was performed by Kolmogorov-Smirnov test. (**c**) Sample recordings of spontaneous EPSCs recorded from pyramidal neurons of non-seizure controls, HS hippocampus, and ATL tissue following perfusion of the slices with action-potential inhibitor TTX (200 nM) for 10 min. (**d**) Frequency of spontaneous EPSCs before and after treatment with TTX. Plot shows data obtained from seven neurons from six non-seizure control patients, six neurons from six ATL, and ten neurons from nine hippocampus of HS patients. *$p < 0.05$, **$p < 0.01$, ***$p < 0.001$ in comparison to the respective controls. Statistical analysis was performed by one-way ANOVA followed by Bonferroni post hoc test. Data is showing as mean and S.E.M

Table 2
Kinetic properties of spontaneous EPSCs (without TTX) and miniature EPSCs (TTX) recorded from pyramidal neurons of non-seizure control tissues

Parameters	Without TTX	TTX
Amplitude (pA)	12.43 ± 1.9	5.9 ± 0.7
Rise time (ms)	2.2 ± 0.7	2.1 ± 0.5
Decay time constant (ms)	9.6 ± 1.4	9.4 ± 1.8

Table 3
Kinetic properties of spontaneous EPSCs (without TTX) and miniature EPSCs (TTX) recorded from pyramidal neurons of HS hippocampus tissues

Parameters	Without TTX	TTX
Amplitude (pA)	15.96 ± 2.1	8.1 ± 1.8
Rise time (ms)	2.0 ± 0.6	2.1 ± 0.5
Decay time constant (ms)	8.3 ± 0.9	9.1 ± 1.8

Table 4
Kinetic properties of spontaneous EPSCs (without TTX) and miniature EPSCs (TTX) recorded from pyramidal neurons of HS ATL tissues

Parameters	Without TTX	TTX
Amplitude (pA)	18.53 ± 2.3	9.4 ± 2.6
Rise time (ms)	2.3 ± 0.6	2.1 ± 0.5
Decay time constant (ms)	9.2 ± 1.3	9.2 ± 2.0

neurotransmission is primarily contributed by glutamatergic tone in the ATL and hippocampus of patients with HS. Glutamatergic excitatory transmission consists of two different components; one is TTX insensitive action potential-independent miniatures (mEPSCs) which manifest quantal release of neurotransmitters, and the other is TTX sensitive action potential-dependent spontaneous EPSCs that indicate neurotransmitter release in response to presynaptic action potentials. The magnitude of TTX-mediated inhibition is higher in ATL than that of the hippocampus, which implies that under resting condition, the action potential-dependent glutamate neurotransmitter release process from presynaptic neurons in ATL is different from that in the hippocampus. So, it can be speculated that pathological glutamatergic neural and synaptic network reorganization accountable for hyperexcitability in ATL is distinctive from that of the hippocampus. These results suggest presence of resting-state, large-scale, independent networks at cellular level in HS patients; one originating from the ATL and others from the hippocampus and excitatory synaptic connections alter significantly between these two regions' consequences in differential glutamatergic responses [19].

3.4 Recording of Glutamatergic Synaptic Activity from Rat TLE

We have recorded spontaneous EPSCs and IPSCs from the pyramidal neurons of CA1 region of hippocampus (Fig. 4a, c). The postsynaptic currents were blocked under influence of respective blockers, viz., EPSCs in presence of APV and CNQX, and IPSCs in presence of bicuculline. The frequencies and amplitudes of EPSCs

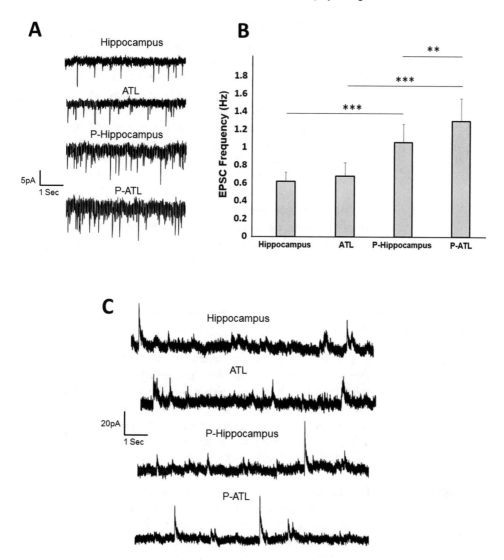

Fig. 4 Spontaneous postsynaptic currents recorded from pyramidal neurons of Sprague-Dawley rat hippo-campus and anterior temporal lobe tissue slices. (**a**) Sample recording of spontaneous EPSCs recorded from pyramidal neurons of hippocampus and ATL obtained from without and with pilocarpine-treated rats. P-Hippocampus and P-ATL depict temporal lobe epilepsy model created by pilocarpine treatment. (**b**) Frequency of EPSCs was significantly higher in the hippocampus and ATL of pilocarpine treated rats as compared to those of untreated animals. However, after pilocarpine treatment frequency of EPSCs in the ATL (P-ATL) samples was also significantly high compared to that in case of the hippocampus (P-Hippocampus). Plot shows data obtained from six neurons from six control rats' hippocampus and ATL; six neurons from six temporal lobe epilepsy rats' hippocampus and ATL. **$p < 0.01$, ***$p < 0.001$ according to one-way ANOVA followed by Bonferroni post hoc test. Data is showing as mean and S.E.M. (**c**) Sample recording of spontaneous IPSCs recorded from pyramidal neurons of hippocampus and ATL obtained from without and with pilocarpine-treated rats

become significantly higher in pilocarpine-treated rats in comparison to the control animals (Fig. 4b). When we record the spontaneous EPSCs from the pyramidal neurons of anterior temporal lobe, we found a region-specific variation of frequencies of spontaneous EPSCs and hence glutamatergic tone between temporal cortex and hippocampus of pilocarpine-treated rats (Fig. 4b). The frequencies of the spontaneous EPSCs recorded from the anterior temporal lobe tissue are significantly higher in comparison to the hippocampus tissues in pilocarpine-treated rats. However, the kinetic properties, that is, rise time (10–90%) and decay time constant, do not alter much after pilocarpine administration in comparison to the control animals (Tables 5, 6, 7, and 8). This also supports the fact that temporal lobe epilepsy is a network disorder, and parallel, independent epileptiform network hubs are present at different areas of temporal cortex contributing toward hyperexcitability.

4   Notes/Troubleshooting

4.1   Viability of Tissue

In case of human brain tissues, all experiments starting from the slice preparation to complete recording of postsynaptic currents need to be completed within 60–70 min. Beyond that point, most of the neurons die out, and data usually becomes unreliable. For rat brain slices, experiments can be performed up to 3 h as neurons are comparatively robust. In both scenario, brain slices should be kept inside well-carbogenated ACSF.

4.2   Chloriding of Electrodes

The reference electrode and recording electrode needs to be checked carefully for proper conductance before every experiment. Chloriding of the silver wire for making electrodes needs to be performed routinely by electroplating.

4.3   Grounding

Every component of a patch clamp rig should be properly connected to earthing ground with copper wires; otherwise electrical noise will interfere with signals.

4.4   Glass Micropipette Preparation

Glass micropipettes should be prepared freshly before every experiment; otherwise its tip can be clogged by dirt particles which prevent it from creating a good patch seal.

4.5   Carbogenation

During the entire course of the experiment starting from slice preparation to entire recording period, extracellular fluid (ACSF) needs to be supplied with carbogen (95% O_2 + 5% CO_2) to keep neurons alive.

4.6   Solutions

ACSF needs to be prepared freshly before every experiment. Internal pipette solution should be aliquoted and kept frozen. Prior to

Table 5
Kinetic properties of spontaneous EPSCs recorded from pyramidal neurons obtained from Sprague-Dawley rat hippocampus tissues without and with pilocarpine treatment (P-Hippocampus; hippocampus of temporal lobe epilepsy model)

Parameters	Hippocampus	P-Hippocampus
Amplitude (pA)	12.58 ± 2.0	16.11 ± 2.2
Rise time (ms)	1.8 ± 0.3	1.9 ± 0.7
Decay time constant (ms)	9.0 ± 1.1	8.8 ± 1.3

Table 6
Kinetic properties of spontaneous EPSCs recorded from pyramidal neurons obtained from Sprague-Dawley rat ATL tissues without and with pilocarpine treatment (P-ATL; anterior temporal lobe of temporal lobe epilepsy model)

Parameters	ATL	P-ATL
Amplitude (pA)	13.19 ± 2.7	18.97 ± 2.9
Rise time (ms)	2.0 ± 0.5	1.9 ± 0.5
Decay time constant (ms)	9.7 ± 1.6	9.9 ± 2.0

Table 7
Kinetic properties of spontaneous IPSCs recorded from pyramidal neurons obtained from Sprague-Dawley rat hippocampus tissues without and with pilocarpine treatment

Parameters	Hippocampus	P-Hippocampus
Frequency (Hz)	1.3 ± 0.2	1.26 ± 0.2
Amplitude (pA)	25.8 ± 0.8	25.2 ± 1.2
Rise time (ms)	2.1 ± 0.18	2.3 ± 0.19
Decay time constant (ms)	36.1 ± 1.9	34.8 ± 1.5

Table 8
Kinetic properties of spontaneous IPSCs recorded from pyramidal neurons obtained from Sprague-Dawley rat hippocampus tissues without and with pilocarpine treatment

Parameters	ATL	P-ATL
Frequency (Hz)	1.4 ± 0.3	1.4 ± 0.3
Amplitude (pA)	26.9 ± 0.8	25.7 ± 0.9
Rise time (ms)	2.0 ± 0.1	2.2 ± 0.1
Decay time constant (ms)	37.3 ± 2.2	37.1 ± 2.6

every experiment, it should be thawed and filtered through 0.22μm syringe filter before use. Stalk concentration of all blockers should be kept frozen, and prior to experiment working concentration should be prepared in ACSF.

4.7 Osmolality of Internal Pipette Solution

Under the whole-cell configuration of patch clamp technique, the internal pipette solution remains in contact with cytoplasm; thus, the osmolarity of the internal pipette solution should be between 290 and 300 mOsm (isosmotic with neuronal cytoplasm). It is recommended to measure the osmolality before every experiment to avoid any detrimental effect on the neuron.

4.8 Access Resistance

During whole-cell configuration of patch clamp recording, the access resistance (pipette resistance plus any current resistive factor at pipette tip) should be between 15–20 MΩ. Any change of more than 20% from this range is not reliable, and the data generated from that neuron should not be considered.

4.9 Leak Current

During whole-cell configuration of patch clamp recording, leak current should be between 50 and 150 pA. If it is more than 200 pA, the data becomes unreliable.

4.10 Peak Selection During Analysis

During analysis of spontaneous EPSCs and IPSCs, for calculating kinetic properties (rise time and decay time constant), events showing sharp peaks and exponential decays should be incorporated. Events showing multiple peaks should not be incorporated for kinetic properties analysis but can be included from frequency calculation.

5 Conclusion

Cellular electrophysiological experiments in the resected brain specimens from the patients with DRE provide the means to investigate the dynamics of synaptic transmission in DRE at single neuron level. Specifically, the combination of this patch clamp technique with individual or combination of pharmacological agents serves as a model system to investigate mechanisms of seizure generation associated with the different pathologies of DRE. Clearly, a significant number of intricate pathological events underlie the creation of epileptogenic network, among which abnormal glutamatergic and GABAergic synaptic transmissions play a noteworthy role. Here we have discussed the use of patch clamp technique to study the alteration in glutamatergic and GABAergic synaptic transmission from a single neuron in the epileptogenic areas of the resected brain specimens obtained from the drug-resistant HS epilepsy patients as well as from TLE animal models. The correlations of cellular electrophysiological data of the epileptogenic network with the pre-surgical multimodal imaging and the clinical data will help enhance our understanding of epileptogenic network. Understanding the epileptogenic networks at circuit level provides a deeper insight into the pathophysiology of DRE.

6 Limitation

The primary limitation of cellular electrophysiological studies on human brain tissue are the non-seizure control which, ideally, should be resected from the temporal cortex areas of patients with similar ages with non-epilepsy pathologies such as glioma. However, it is not possible to get such region-specific tissues due to ethical reason, and in addition to this, non-seizure temporal lobe tumors are remarkably rare. Moreover, temporal cortex obtained from deceased patients undergoing autopsy are not suitable for cellular electrophysiological experiments. Finally, albeit the resected brain specimens serve as an ideal model system to study epileptogenesis, it is important to understand that as these epilepsy patients are on antiepileptic drugs medication, the possible effects of these medicines on synaptic transmission cannot be ruled out. Thus, parallel experiments on animal models will help validate the data obtained from patients to enhance our understanding of epileptogenesis at cellular level.

References

1. Fisher R, Acevedo C, Arzimanoglou A et al (2014) ILAE official report: a practical clinical definition of epilepsy. Epilepsia 55:475–482
2. Murray C, Lopez A (1994) Global comparative assessment in the health sector; disease burden, expenditures, and intervention packages. World Health Organization, Geneva
3. Kwan P, Arzimanoglou A, Berg A et al (2010) Definition of drug resistant epilepsy: consensus proposal by the ad hoc Task Force of the ILAE Commission on Therapeutic Strategies. Epilepsia 51:1069–1077
4. González F, Osorio X, Rein A (2015) Drug-resistant epilepsy: definition and treatment alternatives. Neurologia 30:439–446
5. Dwivedi R, Ramanujam B, Chandra PS et al (2017) Surgery for drug-resistant epilepsy in children. N Engl J Med 377:1639–1647
6. Neher E, Sakmann B (1976) Single-channel currents recorded from membrane of denervated frog muscle fibres. Nature 260:799–802
7. Banerjee J, Chandra PS, Kurwale N et al (2014) Epileptogenic networks and drug-resistant epilepsy: present and future perspectives of epilepsy research-utility for the epileptologist and the epilepsy surgeon. Ann Indian Acad Neurol 17(Supplement 1): S134–S140
8. Aubert S, Wendling F, Regis J et al (2009) Local and remote epileptogenicity in focal cortical dysplasias and neurodevelopmental tumours. Brain 132:3072–3086
9. Firpi H, Smart O, Worrell G et al (2007) High-frequency oscillations detected in epileptic networks using swarmed neural-network features. Ann Biomed Eng 35:1573–1584
10. Elger C, Widman G, Andrzejak R et al (2000) Nonlinear EEG analysis and its potential role in epileptology. Epilepsia 41:S34–S38
11. Fang M, Xi Z, Wu Y et al (2011) A new hypothesis of drug refractory epilepsy: neural network hypothesis. Med Hypotheses 76:871–876
12. Santhakumar V, Aradi I, Soltesz I (2005) Role of mossy fiber sprouting and mossy cell loss in hyperexcitability: a network model of the dentate gyrus incorporating cell types and axonal topography. J Neurophysiol 93:437–453

13. Buckmaster P, Jongen-Rêlo A (1999) Highly specific neuron loss preserves lateral inhibitory circuits in the dentate gyrus of kainate-induced epileptic rats. J Neurosci 19:9519–9529

14. Dixit A, Banerjee B, Tripathi M et al (2015) Presurgical epileptogenic network analysis: a way to enhance epilepsy surgery outcome. Neurol India 63:743–750

15. Tripathi M, Ray S, Chandra PS (2016) Presurgical evaluation for drug refractory epilepsy. Int J Surg 36:405–410

16. Chandra PS, Vaghania G, Bal CS (2014) Role of concordance between ictal-subtracted SPECT and PET in predicting long-term outcomes after epilepsy surgery. Epilepsy Res 108:1782–1789

17. Ramanujam B, Bharti K, Viswanathan V et al (2017) Can ictal-MEG obviate the need for phase II monitoring in people with drug-refractory epilepsy? A prospective observational study. Seizure 45:17–23

18. Tripathi M, Garg A, Gaikwad S et al (2010) Intra-operative electrocorticography in lesional epilepsy. Epilepsy Res 89:133–141

19. Banerjee J, Banerjee Dixit A, Srivastava A et al (2017) Altered glutamatergic tone reveals two distinct resting state networks at the cellular level in hippocampal sclerosis. Sci Rep 7:319

20. Banerjee J, Banerjee Dixit A, Tripathi M et al (2015) Enhanced endogenous activation of NMDA receptors in pyramidal neurons of hippocampal tissues from patients with mesial temporal lobe epilepsy: a mechanism of hyper excitation. Epilepsy Res 117:11–16

21. Dixit AB, Banerjee J, Srivastava A et al (2016) RNA-seq analysis of hippocampal tissues reveals novel candidate genes for drug refractory epilepsy in patients with MTLE-HS. Genomics 107:178–188

22. Curia G, Longo D, Biagini G et al (2008) The pilocarpine model of temporal lobe epilepsy. J Neurosci Methods 172:143–157

23. Racine R (1972) Modification of seizure activity by electrical stimulation. II. Motor seizure. Electroencephalogr Clin Neurophysiol 32:281–294

24. Turski W, Cavalheiro E, Schwarz M et al (1983) Limbic seizures produced by pilocarpine I rats: behavioral, electroencephalographic and neuropathological study. Behav Brain Res 9:315–335

25. Jope R, Morrisett R, Snead O III (1986) Characterization of lithium potentiation of pilocarpine-induced status epilepticus in rats. Exp Neurol 91:471–480

26. Lin JJ, Salamon N, Lee AD et al (2007) Reduced neocortical thickness and complexity mapped in mesial temporal lobe epilepsy with hippocampal sclerosis. Cereb Cortex 17:2007–2018

27. Engel J Jr (1983) Functional localization of epileptogenic lesions. Trends Neurosci 6:60–65

28. Memarian N, Madsen S, Macey P (2015) Ictal depth EEG and MRI structural evidence for two different epileptogenic networks in mesial temporal lobe epilepsy. PLoS One 10: e0123588

29. Cataldi M, Avoli M, deVillers-Sidani E (2013) Resting state networks in temporal lobe epilepsy. Epilepsia 54:2048–2059

30. Jin S, Jeong W, Chung C (2015) Mesial temporal lobe epilepsy with hippocampal sclerosis is a network disorder with altered cortical hubs. Epilepsia 56:772–779

31. Johnston D, Brown TH (1981) Giant synaptic potential hypothesis for epileptiform activity. Science 211:294–297

32. Westbrook G (2013) Seizures and epilepsy. In: Kandel ER, Schwartz J, Jessell T et al (eds) Principles of neural science, 5th edn. The McGraw-Hill, New York, pp 1116–1139

33. Lothman EW (1993) The neurobiology of epileptiform discharges. Am J EEG Technol 33:93–112

34. Josephson C, Dykeman J, Fiest K et al (2013) Systematic review and meta-analysis of standard vs selective temporal lobe epilepsy surgery. Neurology 80:1669–1676

Part III

In Vivo Rodent Models

Chapter 5

Acute Seizure Tests Used in Epilepsy Research: Step-by-Step Protocol of the Maximal Electroshock Seizure (MES) Test, the Maximal Electroshock Seizure Threshold (MEST) Test, and the Pentylenetetrazole (PTZ)-Induced Seizure Test in Rodents

Katarzyna Socała and Piotr Wlaź

Abstract

Maximal electroshock (MES)- and pentylenetetrazole (PTZ)-induced seizures are the two most widely used models of acute seizures in rodents. In the conventional MES test, seizures are induced by applying an electrical stimulus of high intensity and high frequency for a short duration, which evokes tonic seizures that are manifested by a rigid extension of the hindlimbs. The subcutaneous (s.c.) PTZ test is in turn employed to produce clonic seizures. The MES test is considered to be a model of generalized tonic–clonic seizures, whereas the s.c. PTZ-induced seizures are thought to mimic the absence and/or myoclonic epilepsy in humans. These two models are considered to have a high predictive validity for a therapeutic drug response, and both are useful tools in the discovery and development of new antiepileptic drugs. In this chapter, we present the protocols of the traditional MES and s.c. PTZ test in detail. In addition, we describe the method used for the assessment of seizure thresholds in the maximal electroshock seizure threshold (MEST) test, s.c. PTZ test, and the timed intravenous PTZ test in mice. All the protocols include the materials used, step-by-step instructions, and some experimental considerations. Furthermore, the most common problems encountered while performing these procedures are indicated.

Key words Maximal electroshock, Pentylenetetrazole, Seizure threshold, Anticonvulsant, Proconvulsant

1 Introduction

Although a number of animal models of seizures and epilepsy have been established in epilepsy research, two of them, namely, the maximal electroshock (MES) and the subcutaneous (s.c.) pentylenetetrazole (PTZ) seizure tests in rodents, remain the most widely used paradigms for the preliminary assessment of anticonvulsant activity. The MES model has been used for the initial screening of new antiepileptic drugs since the 1930s, when it was employed for

Divya Vohora (ed.), *Experimental and Translational Methods to Screen Drugs Effective Against Seizures and Epilepsy*, Neuromethods, vol. 167, https://doi.org/10.1007/978-1-0716-1254-5_5,

the first time in cats to test novel compounds, which led to the discovery of the anticonvulsant properties of phenytoin [1]. A few years later, the MES model was introduced in rodents by Toman et al. [2]. The use of PTZ as a chemoconvulsant agent for the identification of antiepileptic drugs was started in 1944, when it was demonstrated that the PTZ-induced seizures in mice can be blocked by trimethadione and phenobarbital. Interestingly, phenytoin (which was shown to be effective in the MES test) failed to prevent the seizures induced by PTZ [3]. In addition, trimethadione successfully blocked absence (but not grand mal) seizures in humans [4]. These observations suggested that the MES and s.c. PTZ tests can be used to differentiate antiepileptic drugs with different clinical effects [5, 6]. Accordingly, drugs that are found to be effective in the MES model are thought to prevent generalized tonic–clonic seizures in humans, whereas those active against the s.c. PTZ-induced seizures are considered to be useful in the treatment of absence and/or myoclonic epilepsy in humans. However, it should be noted that in both of these tests, seizures are induced in normal (i.e., nonepileptic) animals. Thus, they do not resemble epilepsy in humans, which is a chronic condition, but represent only acute models of single seizures induced in animals. Despite this limitation, the MES and s.c. PTZ tests have both been well validated and found to be of high predictive value in detecting clinically effective antiepileptic drugs [6–8].

In the MES test in rodents, suprathreshold electrical stimuli are transmitted via transcorneal or less frequently via transauricular electrodes, which induce maximal seizures characterized by tonic extension of the hindlimbs [9] (Fig. 1). The tonic extensor convulsion may be lethal (especially in mice) or lasts for only about 10–15 s, and then it is followed by clonic seizures. The lack of the tonic hindlimb extension (failure of the hindlimbs to extend beyond a 90° angle to the torso in animals) indicates a protective (anticonvulsant) effect. Typically, a stimulus (sine wave or rectangular pulses) with a frequency of 50–60 Hz is applied for a duration of 0.2 s using constant current stimulators [7, 10, 11]. To induce maximal seizures, the current intensity should be at least 2 (preferably 4–5) times higher than the individual seizure threshold of the animals [12]. Such a suprathreshold stimulus avoids the possibility that daily fluctuations in seizure susceptibility affect the induction of tonic seizures [13]. Current intensities of 50 mA and 150 mA are most commonly used for mice and rats, respectively. However, since the seizure threshold varies with the strain, age, weight, and sex of animals, or laboratory conditions, it would be ideal to determine the optimal current intensity in pilot experiments [7, 14–17]. Preliminary studies in rats are especially recommended for this purpose because some strains and old animals may be resistant to the electroshock-induced tonic extension of the hindlimbs [7, 12]. Noteworthy, the threshold current value to elicit tonic

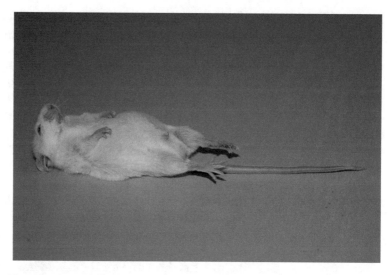

Fig. 1 Fore- and hindlimb tonic extension in the MES/MEST test

seizures by transcorneal stimulation is higher as compared to trans-auricular stimulation, which should also be taken into consideration when planning the experiments [10].

PTZ is the most widely used chemoconvulsant that works as a blocker of the picrotoxin site of the chloride ionophore of the $GABA_A$ receptor complex [18]. The behavioral features of seizures induced by the administration of PTZ differ from those of the seizures observed in the MES test. Depending on the dose used, PTZ induces myoclonic twitches (jerks), minimal clonic seizures (manifested by vibrissae twitching, clonus of the jaw, clonus of the forelimbs and/or hindlimbs without loss of righting reflex), or generalized clonic seizure with loss of righting reflex, and tonic extension of the fore- and hindlimbs [19]. In the s.c. PTZ test, animals are administered with PTZ at the CD_{97} dose (i.e., the dose that produces clonic seizures in 97% of the animals tested) and observed for 30 min. A clonic seizure lasting for at least 5 s is taken as the endpoint. Animals that did not display any clonic seizure in the s.c. PTZ test are considered protected [12, 20]. In mice, PTZ is usually injected at a fixed dose of 85 mg/kg. However, since the seizure thresholds vary depending on many factors, it is recommended to predetermine the CD_{97} dose of PTZ in pilot experiments [19].

For high-throughput screening of novel compounds, animals (3–4 per group) are intraperitoneally (or orally) administered with different doses of the test compound (e.g., 30, 100, and 300 mg/kg) and subjected to the MES and s.c. PTZ test at one or two different time points (e.g., 30 and 240 min). The results of these tests provide preliminary data on the anticonvulsant efficacy of the test compounds. If protection is noted in animals, more detailed time-course studies are performed. Subsequently, an assessment of the dose–response relationship is carried out to determine the ED_{50} value of the compound (i.e., the dose that protects 50% of the animals from seizures). For this purpose, 3–5 groups of animals (6–10 per group) are administered with different doses of the test compound, and the number of animals protected (out of the total number of animals in a group) is noted [19]. The ED_{50} value is then calculated using, for example, the log-probit method described by Litchfield and Wilcoxon [21].

Models with supramaximal seizure induction raise the possibility that some interesting compounds with anticonvulsant properties are overlooked. For this reason, the effect of the test drug on the seizure threshold should be also evaluated. Noteworthy, the compounds that increase seizure threshold not always can provide full protection against the seizures induced by supramaximal stimuli. For example, primidone and clonazepam are known to increase the threshold for tonic hindlimb extension in the maximal electroshock seizure threshold test (MEST), but they are ineffective in the MES test in which seizures are induced by suprathreshold stimuli. Thus, the MES test may yield false-negative results, which may, for example, happen with GABA-enhancing drugs. This can be avoided by parallel use of both the MEST and MES tests [10, 12, 13]. An advantage of the seizure threshold tests is that they allow determining the effect of the drug on the seizure threshold without ignoring individual differences in animals in terms of seizure susceptibility [11]. Moreover, the use of both the seizure threshold test and the maximal or supramaximal model may help in verifying whether the anticonvulsant effect of the test compounds results from their ability to raise the seizure threshold or if other mechanisms, such as suppression of seizure spread, are involved [12]. Seizure threshold tests are also widely employed for assessing the proconvulsant properties of drugs during safety evaluation. Importantly, they can also be used to evaluate the effects of some non-pharmacological factors on seizure susceptibility, such as genetic manipulations, hypo- or hyperthermia, stress, lesions, and so on [7, 19, 22].

In the MEST test, the threshold convulsive current (CC_{50} in mA) causing tonic extension of the hindlimbs in 50% of the animals tested can be determined by using either the "up and down" (also known as "staircase") method described by Kimball et al. [23] or the log-probit method of Litchfield and Wilcoxon [21]. The principles of these two methods are quite the same, but the "up and down" method is faster and requires the use of fewer animals. The threshold PTZ dose required to induce clonic seizures may be determined by the s.c. injection of PTZ at various doses to 3–5 groups of animals and subsequent construction of a dose–response curve to calculate the dose that produces endpoint in 50% of the animals (CD_{50} dose). However, a much more sensitive method is the timed intravenous (i.v.) PTZ seizure threshold test, which allows determining the threshold PTZ dose for several consecutive endpoints [22, 24].

In this chapter, the protocols for the initial screening of new compounds and the determination of the ED_{50} values in the MES and s.c. PTZ tests are presented. Moreover, the methods for assessing thresholds for the MES- and PTZ-induced seizures are described. These procedures are used for mice but require only slight modifications to be used in rats. The most important aspects that should be taken into account when performing the above-mentioned tests and some common problems that are encountered with these procedures are also indicated.

2 Materials

2.1 Materials for the MES and MEST Tests

- Male mice, 20–35 g, housed in groups under standard laboratory conditions. Mice of various strains (e.g., Swiss albino, CD-1, or NMRI) can be used depending on the study design. Females are generally avoided due to the concern that the estrous cycle may induce variability in seizure susceptibility.

- Treatment solutions/suspensions: test compound, vehicle alone (for the control group), and optionally a reference drug (e.g., valproic acid or carbamazepine).

- 0.5% solution of tetracaine hydrochloride (or other topical anesthetics for the eyes).

- Transparent cage (e.g., type III cage for rodents) without bedding.

- Rodent shocker (e.g., type 221; Hugo Sachs Elektronik, Freiburg, Germany) with dedicated corneal electrodes and a foot switch.

- Saline for moistening the electrodes.

- Stopwatches.

2.2 Materials for the s.c. PTZ Test

- Male mice, 20–35 g, housed in groups under standard laboratory conditions. Mice of various strains (e.g., Swiss albino, CD-1, or NMRI) can be used depending on the study design. Females are generally avoided due to the concern that the estrous cycle may induce variability in seizure susceptibility.

- Treatment solutions/suspensions: test compound, vehicle alone (for the control group), and optionally a reference drug (e.g., valproic acid or diazepam).

- PTZ solutions.

- Small transparent cages without bedding.

- Stopwatches.

2.3 Materials for the i.v. PTZ Test

- Male mice, 25–35 g, housed in groups under standard laboratory conditions. Mice of various strains (e.g., Swiss albino, CD-1, or NMRI) can be used. However, it should be noted that the i.v. injection in black strains (e.g., C57BL/6) may be technically difficult because the tail veins poorly contrast with the dark skin. Females are generally avoided due to the concern that the estrous cycle may induce variability in seizure susceptibility.

- Treatment solutions/suspensions: test compound, vehicle alone (for the control group), and optionally a reference drug (e.g., valproic acid or diazepam).

- Mouse restrainer designed for tail vein injections (commercially available or custom made).

- Transparent cage (preferably type III cage for rodents) without bedding.

- Syringe pump (e.g., model Physio 22; Hugo Sachs Elektronik-Harvard Apparatus GmbH, March-Hugstetten, Germany).

- 1% solution of PTZ.

- Ten-milliliter plastic syringe mounted on the pump.

- Two-centimeter-long 27-gauge needles.

- Polyethylene tubing (PE-20, 0.015 × 0.041 in.).

- Stopwatches.

- Optional: adhesive tape, warm water or 70% ethanol and cotton pads.

3 Methods

3.1 Evaluation of the Anticonvulsant Activity in the MES Test

The following procedure is used to evaluate the anticonvulsant activity of the test compounds. Briefly, animals are pretreated with the test compound, and after allowing a suitable period of time for its absorption, they are challenged with the suprathreshold electroshock stimuli. This protocol can be used for both the initial screening of novel antiepileptic drugs and for determining the ED_{50} value of the test compound or antiepileptic drug.

1. Bring mice to the experimental room, and allow them to acclimatize to the environment for at least 30 min. *Do not change the bedding on the day of the experiment. The experiments should be performed during the light phase of the cycle (between 8:00 a.m. and 2:00 p.m.) to minimize the circadian influences.*

2. Prepare the solutions/suspensions of the test compounds and optionally a reference drug for administration in a volume of 10 mL/kg.

3. After acclimatization, weigh the animals and administer the first dose of the test compound (or the reference drug) by intraperitoneal (i.p.), oral (p.o.), or s.c. route depending on the study design. Place an identification mark on the tail of each mouse and immediately return the animal to its home cage.
   For the initial screening in the MES test, 3–4 mice are sufficient for a group. However, for determining the ED_{50} value, there should be at least 6 mice in each group.

4. About 1–2 min before stimulation, apply a drop of ocular anesthetic on the eyes of each mouse to reduce corneal pain.

5. Set the rodent shocker apparatus to deliver a 50-mA stimulus at a frequency of 50–60 Hz for a duration of 0.2 s.
   Usually, the current intensity used in this test is 4–5 times higher than the seizure threshold established in the MEST test.

6. Soak the electrodes with saline to facilitate good electrical contact with the corneas.

7. When the posttreatment time has elapsed, begin testing the mice. Gently grasp each animal by the scruff of the neck behind the ears, and place the saline-soaked corneal electrodes directly on the eyes. Press a foot switch to trigger the stimulus.
   It is important to allow good electrical contact between the eyes and the electrodes. If the contact between the cornea and the electrode circuit is unreliable, some models of electroshock stimulators (including type 221 Rodent Shocker, Hugo Sachs Elektronik) do not deliver the stimulus, which is signaled by an acoustic and display signal. In that event, correct the placement of electrodes and trigger the stimulus again. Moreover, sparks occur (very rarely) indicating that the current passed through the air between the electrode and the cornea. In that case, the mouse may not have received the intended stimulus, and therefore, it should be eliminated from the study (see **Note 1***).*

8. Immediately after the stimulation, place the animal into a transparent cage for behavioral observation.

9. Record the presence or absence of the endpoint (i.e., the tonic hindlimb extension, which is defined as the rigid extension of the hindlimb that exceeds a 90° angle with the body, Fig. 1).
   Additionally, the duration of the tonic phase, flexion/extension ratio, and occurrence of death may be recorded depending on the study design.

10. Repeat **steps 6–9** for all the animals in the experimental group. The animals should be tested in the order they were administered with the test compound.

11. Immediately euthanize all the surviving animals.

Table 1
Evaluation of the ED_{50} value of carbamazepine (a reference compound) in the MES test in mice

Treatment	Number of mice protected from the tonic hindlimb extension	Number of animals in the group
Carbamazepine 10 mg/kg	1	8
Carbamazepine 11 mg/kg	3	8
Carbamazepine 12 mg/kg	5	8
Carbamazepine 14 mg/kg	7	8
	ED_{50} (95% CI) = 11.66 (10.53–12.91) mg/kg $N = 16$	

Carbamazepine was injected i.p. 30 min before the test. ED_{50}, median effective dose; 95% CI, 95% confidence interval; N, number of animals between ED_{16} and ED_{84}, i.e., the number of animals at those doses, whose anticonvulsant effects ranged between 16% and 84% (this value is used for statistical analysis)

12. Express the results as the number of mice that failed to show hindlimb extension out of the total number of mice tested in the experimental group.

13. Repeat the whole protocol for the next experimental group administered with the test compound at a higher or lower dose depending on the percentage of mice protected from seizures recorded in the previous group. Calculate the ED_{50} value of the test compound (i.e., the dose (in mg/kg) that protects 50% of the animals from the tonic hindlimb extension) with the corresponding 95% confidence limits using the log-probit method described by Litchfield and Wilcoxon [21].
    Note that for the determination of the ED_{50} value, 3–5 subgroups of animals (6–10 per subgroup) are required. Each subgroup is administered with a different dose of a test compound or a reference drug (sample results are shown in Table 1). For the statistical comparison of the ED_{50} values determined for two or more groups, use Student's t-test or analysis of variance (ANOVA), where appropriate. Remember that the sample size is the number of animals at those doses, whose anticonvulsant effects ranged between 16% and 84% (not the total number of animals used to determine the ED_{50} value).

3.2 Determination of the Seizure Threshold in the MEST Test in Mice Using the Log-Probit Method

This model can be used to evaluate the pro- or anticonvulsant effects of the test compounds. In addition, the influence of non-pharmacological factors on the seizure threshold can be evaluated by this method. To determine the electroconvulsive threshold using the log-probit method, groups of mice are challenged with electroshocks of variable intensity, and the percentage of animals exhibiting seizure (tonic extension of the hindlimbs) is noted.

1. Bring mice to the experimental room and allow them to acclimatize to the environment for at least 30 min.

   Do not change the bedding on the day of the experiment. The experiments should be performed during the light phase of the cycle (between 8:00 a.m. and 2:00 p.m.) to minimize the circadian influences.

2. Prepare the solutions/suspensions of the test compounds, vehicle (for the control group), and optionally a reference drug for administration in a volume of 10 mL/kg.

3. After acclimatization, weigh the animals (at least 6) and administer vehicle, a reference drug, or the test compound at the first dose by i.p., p.o., or s.c. route depending on the study design. Place an identification mark on the tail of each mouse and immediately return the animal to its home cage.

4. About 1–2 min before stimulation, apply a drop of ocular anesthetic on the eyes of each mouse to reduce corneal pain.

5. Set the rodent shocker apparatus to deliver an 8-mA stimulus at a frequency of 50–60 Hz for a duration of 0.2 s.

   For the vehicle-treated group, the initial current intensity should be close to the threshold value determined in the other experiment. When the ear-clip electrodes are used, the first animal should be challenged with an initial stimulus of lower intensity (between 4 and 6 mA). Moreover, for the compound- or the reference drug-treated group, the initial current strength should be lower or higher than that used for the vehicle-treated group depending on the expected results.

6. Soak the electrodes with saline to allow good electrical contact with the corneas (*see* **Note 1**).

7. When the posttreatment time has elapsed, begin testing the mice. Gently grasp each animal by the scruff of the neck behind the ears, and place the saline-soaked corneal electrodes directly on the eyes. Press a foot switch to trigger the stimulus.

   Remember to ensure good electrical contact between the eyes and the electrodes, as described earlier (in Subheading 3.1, step 7).

8. Immediately after the stimulation, place the animal into a transparent cage for behavioral observation.

9. Record the presence or absence of the tonic hindlimb extension (Fig. 1).

10. Immediately euthanize all the surviving animals.

11. Repeat **steps 6–10** for all the mice in the experimental group.

12. Depending on the results obtained for the previous group, repeat **steps 5–11** for the other groups of animals changing only the current intensity.

88 Katarzyna Socała and Piotr Wlaź

The current intensity should be changed by the same value using a log scale (e.g., 0.06-log intervals as described below in the "up and down" method) or a multiplier (e.g., of 1.2).

13. Using the log-probit method described by Litchfield and Wilcoxon [21], calculate the seizure threshold which is defined as the median current strength (CC_{50} in mA) representing the current intensity that is necessary to induce tonic hindlimb extension in 50% of the animals tested.
    Note that for estimating the CC_{50} value in one experimental group (i.e., test compound-, reference drug-, or vehicle-treated group), 3–5 subgroups of animals (6–10 per subgroup) are required. Each subgroup is challenged with a different current intensity.

14. Repeat the whole protocol for each experimental group.
    For the statistical comparison of the CC_{50} values determined for two or more groups, use Student's t-test or ANOVA (with the post hoc test) where appropriate. Remember that the sample size is the number of animals at those current strength intensities whose seizure effects ranged between 16% and 84% (not the total number of animals used to determine the CC_{50} value).

3.3 Determination of the Seizure Threshold in the MEST Test in Mice Using the "Up and Down" Method

Similar to the aforementioned procedure, the "up and down" method can be used to determine the pro- and anticonvulsant effects of the test compounds or some nonpharmacological factors. To evaluate the electroconvulsive threshold using this method, subsequent mice are challenged with electroshocks of lower or higher intensity depending on whether the previously stimulated animal did or did not exhibit tonic hindlimb extension, respectively.

1. Bring mice to the experimental room and allow them to acclimatize to the environment for at least 30 min.
   Do not change the bedding on the day of the experiment. The experiments should be performed during the light phase of the cycle (between 8:00 a.m. and 2:00 p.m.) to minimize the circadian influences.

2. Prepare the solutions/suspensions of the test compounds, vehicle (for the control group), and optionally a reference drug for administration in a volume of 10 mL/kg.

3. After acclimatization, weigh the animals (15–25 per group) and administer vehicle, a reference drug, or the test compound at the first dose by i.p., p.o., or s.c. route depending on the study design. Place an identification mark on the tail of each mouse and immediately return the animal to its home cage.

4. About 1–2 min before stimulation, apply a drop of ocular anesthetic on the eyes of each mouse to reduce corneal pain.

5. Set the rodent shocker apparatus to deliver an initial stimulus of 10 mA at a frequency of 50–60 Hz for a duration of 0.2 s.

For the vehicle-treated group, the initial current intensity should be close to the threshold value determined in the other experiments. It may vary depending on the strain, age, and sex of mice used in the study or the experimental conditions. If the ear-clip electrodes are used, the first animal should be challenged with an initial stimulus of lower intensity (e.g., 4–6 mA). Moreover, for the compound- or a reference drug-treated group, the initial current strength should be lower or higher than that used for the vehicle-treated group depending on the expected results.

6. Soak the electrodes (usually made from stainless steel or copper covered by thin leather or cotton) with saline to allow good electrical contact with the corneas.

7. When the posttreatment time has elapsed, begin testing the mice. Gently grasp each animal by the scruff of the neck behind the ears, and place the saline-soaked corneal electrodes directly on the eyes. Press a foot switch to trigger the stimulus.
Remember to ensure good electrical contact between the eyes and the electrodes, as described earlier (in Subheading 3.1, step 7).

8. Immediately after the stimulation, place the animal into a transparent cage for behavioral observation.

9. Record the presence or absence of the tonic hindlimb extension (Fig. 1).

10. For the next mouse, repeat **steps 5–9** changing only the current intensity. If the first animal responded with tonic hindlimb extension, the next one should be stimulated with the current of an intensity that is 0.06-log step lower than the previous one. However, if the first mouse did not respond with convulsions, the next one should be stimulated with the current of an intensity that is 0.06-log step higher than the previous one.

11. Repeat **step 10** until all mice from the experimental group are tested. The animals should be tested in the order they were administered with the test compound or vehicle.

12. Euthanize all the surviving animals.

13. Calculate the threshold current causing endpoint in 50% of the mice tested (i.e., CC_{50} value with confidence limits for 95% probability) for each experimental group using the method described by Kimball et al. [23] (sample results are shown in Table 2).

14. Repeat the whole protocol for the next experimental group.
Compare the obtained results using Student's t-test or ANOVA (with the post hoc *test) where appropriate. Note that for the statistical comparison of the CC_{50} values, the sample size should be the number of animals that showed the less frequent outcome and not the total number of animals used.*

Table 2
Effect of a valproic acid (a reference drug) on the seizure threshold in the MEST test in mice performed according to the "up and down" method described by Kimball et al. [23]

Treatment	Current intensity (mA)	Current intensity (log mA)	Mouse number																				
			1	2	3	4	5	6	7	8	9	10	11	12	13	14	15	16	17	18	19	20
Saline, i.p. 10 mL/kg 15 min before the test	7.6	0.88	+						–										–			–
	8.7	0.94		–	+	–	+	+		–	–		+		–	–		+	–		+	
	10.0	1.00									+	+	+	+	+	+	+	+		+		
	11.5	1.06																				
			CC$_{50}$ (95% CI) = 9.20 (8.80–9.63) mA Mean of log mA ± SD = 0.964 ± 0.031 N = 10																				
Valproic acid i.p. 150 mg/kg 15 min before the test	11.5	1.06		–												–		–				–
	13.2	1.12				+		–			–		–			+	–		–			
	15.1	1.18						+	+	+	+			+	+		+		+	+		
	17.4	1.24												+								
			CC$_{50}$ (95% CI) = 14.57 (13.65–15.55) mA Mean of log mA ± SD = 1.163 ± 0.042*** N = 9																				

CC$_{50}$, median convulsive current (in mA) required to produce tonic hindlimb extension in 50% of the tested animals; 95% CI, 95% confidence interval, N number of mice showing the less frequent outcome (this value is used for statistical analysis); *** $p < 0.001$ vs. the vehicle-treated group (Student's t-test)

**3.4 Determination
of the Anticonvulsant
Activity in the s.c. PTZ
Test in Mice**

As the MES test, the following procedure is used for evaluating the anticonvulsant activity of the test compounds. Briefly, animals are pretreated with the test compound, and after allowing a suitable period of time for absorption, they are injected with a convulsive dose of PTZ. The protocol can be used for the initial screening of novel antiepileptic drugs as well as for further characterization of the anticonvulsant potential of the test compound by determining its ED_{50} value.

1. Bring mice to the experimental room and allow them to acclimatize to the environment for at least 30 min.
   Do not change the bedding on the day of the experiment. The experiments should be performed during the light phase of the cycle (between 8:00 a.m. and 2:00 p.m.) to minimize the circadian influences.

2. Prepare the solutions/suspensions of the test compounds and optionally a reference drug for administration in a volume of 10 mL/kg.

3. After acclimatization, weigh the animals (at least 6), and administer the first dose of the test compound or reference drug by i.p., p.o., or s.c. route depending on the study protocol. Place an identification mark on the tail of each mouse, and immediately return the animal to its home cage.

4. Prepare the PTZ solution for administration in a volume of 10 mL/kg, at the predetermined CD_{97} dose (or alternatively at a dose of 85 mg/kg).

5. When the posttreatment time has elapsed, begin testing the mice. Gently grasp each animal, and inject the PTZ solution into a loose fold of the skin in the midline of its neck.
   To minimize leakage, the needle should be inserted several millimeters through the s.c. tissue.

6. Immediately following the administration of PTZ, place the mouse into a transparent cage for behavioral observation. Start the stopwatch.

7. Repeat **steps 5** and **6** for all the animals in the experimental group. The animals should be tested in the order they were administered with the test compound.

8. Observe the animals for 30 min for the occurrence of clonic seizures lasting for at least 3 s.

9. Record the number of mice exhibiting clonic seizures out of the total number of mice tested in the group.
   Additionally, other parameters, including latency to the first myoclonic twitch, generalized clonus and fore- and/or hindlimb

tonus, number of myoclonic, clonic, and tonic seizures, and number of deaths, can be recorded depending on the study design.

10. After observing for 30 min, euthanize all the surviving animals.

11. Express the results as the number of mice that failed to show the endpoint out of the total number of mice tested in the experimental group.

12. If the ED_{50} value is determined, repeat the whole protocol for the next experimental group administered with the test compound (or the reference drug) at a higher or lower dose depending on the percentage of mice exhibiting clonic seizures recorded in the previous group. Using the log-probit method described by Litchfield and Wilcoxon [21], calculate the ED_{50} value of the test compound or the reference drug (i.e., the dose (in mg/kg) that protects 50% of the animals from the clonic seizures induced by s.c. PTZ) with corresponding 95% confidence limits.

Note that for estimating the ED_{50} value, 3–5 subgroups of animals (6–10 per subgroup) are required. Each subgroup is administered with a different dose of the test compound or the reference drug. For the statistical comparison of the ED_{50} values determined for two or more groups, use Student's t-test or ANOVA (with the post hoc *test) where appropriate. Remember that the sample size is the number of animals at those doses, whose anticonvulsant effects ranging between 16% and 84% (not the total number of animals used to determine the ED_{50} values).*

3.5 Determination of the Seizure Threshold in the s.c. PTZ Test in Mice

This procedure is used to evaluate the pro- and anticonvulsant effects of the test compounds or other factors. To determine the threshold for s.c. PTZ-induced seizures, groups of mice are injected with different doses of PTZ, and the percentage of mice exhibiting clonic seizures is recorded.

1. Bring mice to the experimental room, and allow them to acclimatize to the environment for at least 30 min.

Do not change the bedding on the day of the experiment. The experiments should be performed during the light phase of the cycle (between 8:00 a.m. and 2:00 p.m.) to minimize the circadian influences.

2. Prepare the solutions/suspensions of the test compounds, vehicle (for the control group), and optionally a reference drug for administration in a volume of 10 mL/kg.

3. After acclimatization, weigh the animals (at least 6) and administer the test compound, vehicle, or a reference drug by i.p., p. o., or s.c. route depending on the study protocol. Place an

identification mark on the tail of each mouse, and immediately return the animal to its home cage.

4. Prepare 3–5 solutions of PTZ at doses ranging from 50 to 120 mg/kg (injection volume is 10 mL/kg).

5. When the posttreatment time has elapsed, begin testing the animals. Gently grasp each animal, and inject PTZ solution (at the first dose) into a loose fold of skin in the midline of its neck.

   To minimize leakage, the needle should be inserted several millimeters through the s.c. tissue.

6. Immediately following the administration of PTZ, place the mouse into a transparent cage for behavioral observation. Start the stopwatch.

7. Repeat **steps 5** and **6** for all the animals in the experimental group. The animals should be tested in the order they were administered with the test compound.

8. Observe the animals for 30 min for the occurrence of clonic seizures lasting for at least 3 s.

9. Record the number of mice exhibiting clonic seizures out of the total number of mice tested in the group.

10. After observing for 30 min, euthanize all the animals.

11. Repeat **steps 3–10** for the next group administered with PTZ at a higher or lower dose depending on the percentage of mice exhibiting clonic seizures recorded in the previous group.

12. Using the log-probit method described by Litchfield and Wilcoxon [21], calculate the median convulsive dose (CD_{50}) of PTZ (i.e., the dose (in mg/kg) that induces clonic seizures in 50% of the animals) with corresponding 95% confidence limits.

    Note that for estimating the CD_{50} value in one experimental group (i.e., test compound-, reference drug-, or vehicle-treated group), 3–5 subgroups of animals (6–10 per subgroup) are required. Each subgroup is challenged with PTZ at a different dose.

    In addition, based on the results observed for the vehicle-treated group, the CD_{97} dose of PTZ (i.e., the dose of PTZ (in mg/kg) that produces clonic seizures in 97% of the mice tested) can be calculated using the log-probit method [21] (sample results are shown in Table 3*) (see* **Note 2***).*

13. Repeat the whole protocol for the next experimental group.

    For the statistical comparison of the CD_{50} values determined for two or more groups, use Student's t-test or ANOVA (with the post hoc *test) where appropriate. Remember that the sample size is the number of animals at those PTZ doses whose seizure effects ranging between 16% and 84% (not the total number of animals used to determine the CD_{50} value).*

Table 3
Evaluation of the CD_{50} and CD_{97} value of PTZ in the s.c. PTZ test in the vehicle-treated mice

Treatment	Number of animals with clonic seizure	Number of animals in the group
PTZ 60 mg/kg	2	8
PTZ 72 mg/kg	6	8
PTZ 86 mg/kg	7	8
	CD_{50} (95% CI) = 66.66 (58.21–76.34) mg/kg $N = 16$ $CD_{97} = 95.90$ mg/kg	

CD_{50}, median convulsive dose (i.e., PTZ dose producing clonic seizures in 50% of animals); 95% CI, 95% confidence interval; N number of animals between ED_{16} and ED_{84}, i.e., the number of animals at those PTZ doses, whose proconvulsant effects ranged between 16% and 84% (this value is used for statistical analysis); CD_{97}, PTZ dose producing clonic seizures in 97% of animals

3.6 Determination of the Seizure Threshold in the Timed i.v. PTZ Test in Mice

This is an extremely sensitive method for assessing the threshold for the PTZ-induced seizures and allows detecting both the anti- and proconvulsant effects of the test compounds or other factors. Briefly, animals are infused with PTZ solution, and the latency to three consecutive endpoints is measured. Then, the threshold dose of PTZ for each endpoint is calculated.

1. Bring mice to the experimental room and allow them to acclimatize to the environment for at least 30 min.
   Do not change the bedding on the day of the experiment. The experiments should be performed during the light phase of the cycle (between 8:00 a.m. and 2:00 p.m.) to minimize the circadian influences.

2. Prepare the solutions/suspensions of the test compounds for administration in a volume of 10 mL/kg.

3. For the i.v. injection, prepare a 1% solution of PTZ in saline. Fill a 10-mL syringe with the PTZ solution and load the syringe into the pump. The syringe should be connected by polyethylene tubing to a 27-gauge needle. Make sure that there are no air bubbles in the tubing.

4. After acclimatization, weigh 10–12 mice (record the body weights of each) and administer vehicle, a reference drug, or the test compound at the first dose by i.p., p.o., or s.c. route depending on the study design. Place an identification mark on the tail of each mouse and immediately return the animal to its home cage. Stagger the administration to maintain the same time between dosing and testing for each mouse.
   The time intervals used to stagger the administration depend on the time taken to test each animal. If a 1% solution of PTZ is infused at a rate of 0.2 mL/min, 5-min intervals are sufficient (see Note 3).

5. When the posttreatment time has elapsed, begin testing the animals. Place the mouse in a restrainer of an appropriate size. To induce peripheral vasodilatation and increase the visibility of the veins, the tail of the animal can be warmed by dipping in warm water (40–45 °C) or swabbing with 70% alcohol.

6. Gently grasp the distal part of the tail (below the injection site) and slightly rotate to access the lateral tail vein. Insert the needle up to 10 mm into the vein. Secure the needle to the tail by a piece of adhesive tape (optional).

   The accuracy of needle placement in the vein is confirmed by the appearance of blood in the tubing. The tail can be slightly pressed with a thumb (above the injection site) to check if the blood is moving through the tubing. If there is no blood in the tubing or if the blood is not moving, the needle may not be in the vein. In that case, withdraw the needle and repeat the injection. The first attempt of injection should be made approximately in the middle or in the distal one-third of the tail so that the subsequent attempts can be done more proximally (i.e., at a site closer to the body). If the i.v. puncture is still unsuccessful, the second lateral vein can be used. After a few failed attempts, the animal must be eliminated from the study (see Note 3). Furthermore, the duration of the restraint should be kept to a minimum.

7. Upon successful needle placement, release the mouse from the restrainer and gently put it into an empty transparent cage for behavioral observation. The animal should move freely in the cage without strain on the attached polyethylene tubing or struggling.

8. After releasing the animal, start the stopwatch simultaneously with the infusive pump and record the time intervals (in seconds) from the beginning of the PTZ infusion to the onset of the following endpoints: the first myoclonic twitch, generalized clonus with loss of the righting reflex, and tonic fore- and/or hindlimb extension (Figs. 2 and 3).

   The needle may come out of the vein during the infusion. One of the signs of needle displacement, especially at the beginning of PTZ infusion, is when the mouse licks its tail at the injection site and/or flicks the tail. The second sign is the tail becoming blanched around the injection site. If this happens, stop the infusion and eliminate the animal from the experiment.

9. PTZ infusion should be terminated at the appearance of the last recorded endpoint. Tonic seizures are usually lethal for mice. Immediately after observation, all the surviving animals should be euthanized.

10. In the case of nonappearance of seizure(s), but when the needle is inserted correctly, continue the infusion until the predetermined maximum time is reached. The time at which the

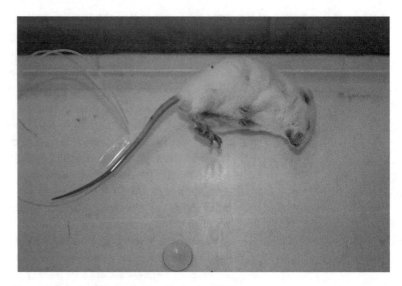

Fig. 2 Forelimb tonic extension in the i.v. PTZ test

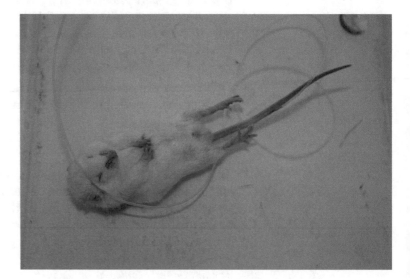

Fig. 3 Fore- and hindlimb tonic extension in the i.v. PTZ test

infusion should be stopped depends on the infusion rate and the study design.

11. Repeat **steps 5–10** for each mouse in the experimental groups.

12. Calculate the seizure thresholds (i.e., the amount of PTZ (in mg/kg) needed to produce the first apparent sign of each endpoint), separately for each endpoint, using the following formula:

Threshold dose of PTZ (mg/kg)

$$= \frac{\text{infusion time (s)} \times \text{infusion rate (mL/s)} \times \text{PTZ concentration (mg/mL)}}{\text{body weight (kg)}}$$

13. Repeat the whole protocol for the next experimental group.

14. Data should be expressed as mean $\pm$ SEM for each group and analyzed using Student's t-test or ANOVA (followed by the post hoc test) to detect statistically significant differences in the seizure thresholds. Sample results are shown in Table 4. For general considerations of planning experiments in Sect 3.1–3.6, please refer to **Notes 4–7**.

4   Notes

All the protocols described in this chapter are well standardized, easy to perform (after adequate training), and require only initial investments in the equipment. Depending on the model used, the seizure phenotype can vary. Therefore, experience is needed to differentiate between the types of seizures and avoid recording an inappropriate endpoint. From a technical viewpoint, the MES/MEST and s.c. PTZ tests are simple to perform and do not involve major problems.

1. A common fault that is noticed with the MES and MEST tests is poor electrical contact between the electrodes and the site of stimulation (i.e., the corneas or less frequently the ear lobes). Practice is required to restrain the animals properly and ensure the correct placement of the electrodes.

2. In the case of the s.c. PTZ test, a problem is that the fixed dose of 85 mg/kg may be too low or too high. For this reason, it is recommended to perform a preliminary study to determine the convulsive dose of PTZ (CD_{97}).

3. Undoubtedly, the timed i.v. PTZ seizure threshold test requires a high level of skill and technical expertise. The most common problems encountered with this procedure are unsuccessful catheterization and needle displacement during PTZ infusion. A well-trained experimenter can achieve a successful i.v. injection rate of about 80–90%, but appropriate training must be undertaken before performing an actual experiment. The next problem that is faced during the i.v. PTZ infusion is that the seizures may develop rapidly with very short intervals between the endpoints. This happens when the PTZ concentration is too high and/or if the infusion rate is too fast. By contrast, a very low PTZ dose and/or infusion rate will cause the seizures to develop very late with long intervals between

Table 4
Effect of a valproic acid (a reference drug) on the seizure threshold in the timed i.v. PTZ test in mice

Treatment	Mouse No.	Body weight (g)	Initial myoclonic twitch		Generalized clonus with loss of righting reflex		Forelimb tonus	
			Infusion time (s)	PTZ (mg/kg)	Infusion time (s)	PTZ (mg/kg)	Infusion time (s)	PTZ (mg/kg)
Saline, i.p. 10 mL/kg 15 min before the test	1	26	36	46.1	40	51.3	97	124.3
	2	26	27	34.6	28	35.9	73	93.6
	3	29	36	41.4	40	46.0	63	72.4
	4	31	38	40.9	42	45.2	94	101.1
	5	30	42	46.7	46	51.1	93	103.3
	6	26	28	35.9	30	38.5	58	74.4
	7	32	39	40.6	50	52.1	65	67.7
	8	33	37	37.4	50	50.5	102	103.0
	9	28	34	40.5	44	52.4	48	57.1
	10	30	37	41.1	43	47.8	50	55.6
	Mean ± SEM =			40.5 ± 1.2		47.1 ± 1.8		85.3 ± 7.3
Valproic acid, i.p. 150 mg/kg 15 min before the test	1	28	49	58.3	53	63.1	120	142.8
	2	26	49	62.8	51	65.4	114	146.1
	3	30	60	66.7	73	81.1	147	163.3
	4	27	45	55.6	47	58.0	109	134.6
	5	26	37	47.4	39	50.0	117	150.0
	6	26	45	57.7	48	61.5	117	150.0
	7	26	44	56.4	45	57.7	100	128.2
	8	26	45	57.7	46	59.0	120	153.8
	9	27	34	42.0	43	53.1	87	107.4
	10	29	45	51.7	49	56.3	109	125.3
	Mean ± SEM =			55.6 ± 2.3***		60.5 ± 2.7***		140.2 ± 5.2***

PTZ (1% solution) was infused at a rate of 0.2 mL/min, ***$p < 0.001$ vs. the vehicle-treated group (Student's t-test) the endpoints. The concentration of PTZ recommended for mice and rats is 1% and 0.8%, respectively. The infusion rate of PTZ for mice varies from 0.05 to 0.5 mL/min, but the optimum rate of infusion is 0.2–0.3 mL/min. In rats, PTZ is typically infused at a rate of 1 mL/min. The parameters of the i.v. PTZ seizure threshold test (i.e., PTZ concentration and infusion rate) should be optimized in each laboratory depending on the study design, for example, so that both the pro- and anticonvulsant effects of the test compound could be evaluated using the same parameters.

4. A number of factors may influence seizure susceptibility and should be taken into consideration while planning and performing the experiments. Not only the strain, age, and sex of the animals but also the husbandry conditions (temperature, light/dark cycle, etc.) and stress may affect the seizure threshold [7, 14–17, 25]. Therefore, a minimum period of 7 days should be allowed for the animals to acclimatize to the laboratory conditions before they are used in seizure tests. All the experiments should be performed at the same time of day to minimize circadian influences. Moreover, control and drug experiments must be always done on the same day to avoid the influence of day-to-day variations on seizure threshold [12, 13]. Other important factors that influence seizure susceptibility include strain, age, sex, and hormonal changes of the animals [7, 14, 25]. For this reason, it is recommended to perform preliminary experiments to determine the threshold for electrically and/or chemically induced seizures and to set up optimal parameters for the actual experiments.

5. All efforts should be made to minimize the suffering of animals during these procedures. As mentioned above, tonic extension of the hindlimbs can be lethal, especially for mice. The surviving animals should be euthanized immediately following the observation period. In acute seizure tests, animals should be used only once unless the experiment is designed otherwise. If the animal is used more than once, the postictal increase in seizure threshold may occur [12], which is another matter of consideration.

6. The timescale required to complete the described protocols depends on several factors, including the procedure used, number of treatment groups, group size, pretreatment times, and number of time points to be tested. The time taken to prepare the equipment and the required substances, the period allowed for the animals to acclimatize to the experimental room, and staggering of the administration of test compounds should also be taken into account when planning the experiment.

7. As regards the sample size, the number of animals to be used in these studies must be justified and reasonable. For mass screening of novel compounds in the MES or the s.c. PTZ test, 3–4 animals per dose (or time point) is fair enough. In the i.v. PTZ seizure threshold test, it is recommended to use 10–12 animals per group because about 10–20% of them are expected to be excluded from the study due to unsuccessful catheterization or needle displacement during PTZ infusion. To determine the seizure threshold in the MEST test using the "up and down" method, 15–25 animals per group (optimally 20) are needed. If the log-probit method is used to determine the seizure threshold in the MEST (CC_{50} value) and the s.c. PTZ (CD_{50} value) test or to calculate the ED_{50} doses of the test compounds, at least three groups of animals are required to observe a dose–response effect. The response should be close to 0%, 50%, and 100%. One or two additional groups of animals can be used if necessary, and the number of animals per group may vary from 6 to 10 (usually 8).

Tables 1, 2, 3, and 4 show the example data obtained in the MES, MEST, and PTZ-induced seizure tests. Depending on the procedure used and the study design, the results can be analyzed using different statistical tools (typically Student's *t*-test or ANOVA). However, it should be noted that for the statistical comparison of the ED_{50} or CD_{50} values, the sample size should be the number of animals at those doses whose response ranged between 16% and 84% (not the total number of animals used to determine the ED_{50} or CD_{50} value). Likewise, the sample size for the statistical comparison of the CC_{50} values determined using the log-probit method should be the number of animals at those current strengths whose response ranged between 16% and 84%. When the "up and down" method is used, the number of animals that showed the less frequent response should be considered for statistical analysis, and not the total number of animals used.

5   Conclusions

The MES and s.c. PTZ tests are simple screening tools that have provided an estimate for the potential anticonvulsant efficacy of compounds in the early stage of drug discovery for a few decades. Both these tests have contributed to the introduction of several clinically effective antiepileptic drugs. Seizure threshold tests allow identification of anticonvulsant properties of test compounds or other factors. In addition, they also provide data on possible pro-convulsant effects. All the procedures described in this chapter are technically easy to perform, are not time-consuming, and give reproducible results. However, they have several limitations, which are widely discussed elsewhere. One of the main concerns is that these procedures cannot identify the drugs that can be useful in the treatment of refractory epilepsy. Nevertheless, the aforementioned acute seizure tests are still widely used in epilepsy research.

References

1. Putnam TJ, Merritt HH (1937) Experimental determination of the anticonvulsant properties of some phenyl derivatives. Science 85:525–526

2. Toman JE, Swinyard EA, Goodman LS (1946) Some properties of maximal electroshock seizures. Fed Proc 5:105

3. Everett GM, Richards RK (1944) Comparative anticonvulsive action of 3,5,5-trimethyloxazo-lidine-2,4-dione (Tridione), dilantin and phenobarbital. J Pharmacol Exp Ther 81:402–407

4. Lennox WG (1945) The petit mal epilepsies; their treatment with tridione. J Am Med Assoc 129:1069–1074

5. White HS, Johnson M, Wolf HH, Kupferberg HJ (1995) The early identification of anticonvulsant activity: role of the maximal electroshock and subcutaneous pentylenetetrazol seizure models. Ital J Neurol Sci 16:73–77

6. Löscher W (2017) Animal models of seizures and epilepsy: past, present, and future role for the discovery of antiseizure drugs. Neurochem Res 42:1873–1888

7. Giardina WJ, Gasior M (2009) Acute seizure tests in epilepsy research: electroshock- and chemical-induced convulsions in the mouse. Curr Protoc Pharmacol 45:5.22.1–5.22.37

8. Löscher W (2016) Fit for purpose application of currently existing animal models in the discovery of novel epilepsy therapies. Epilepsy Res 126:157–184

9. Wlaź P, Potschka H, Löscher W (1998) Frontal versus transcorneal stimulation to induce maximal electroshock seizures or kindling in mice and rats. Epilepsy Res 30:219–229

10. Peterson SL (1998) Electroshock. In: Peterson SL, Albertson TE (eds) Neuropharmacology methods in epilepsy research. CRC Press, Boca Raton, pp 1–26

11. van Vliet E, Gorter J (2017) Electrical stimulation seizure models. In: Pitkänen A, Buckmaster PS, Galanopoulou AS, Moshe SL (eds) Models of seizures and epilepsy, 2nd edn. Elsevier Inc, London, pp 475–491

12. Löscher W, Schmidt D (1988) Which animal models should be used in the search for new antiepileptic drugs? A proposal based on experimental and clinical considerations. Epilepsy Res 2:145–181

13. Castel-Branco MM, Alves GL, Figueiredo IV, Falcao AC, Caramona MM (2009) The maximal electroshock seizure (MES) model in the preclinical assessment of potential new antiepileptic drugs. Methods Find Exp Clin Pharmacol 31:101–106

14. Frankel WN, Taylor L, Beyer B, Tempel BL, White HS (2001) Electroconvulsive thresholds of inbred mouse strains. Genomics 74:306–312

15. Ferraro TN, Golden GT, Smith GG, DeMuth D, Buono RJ, Berrettini WH (2002) Mouse strain variation in maximal electroshock seizure threshold. Brain Res 936:82–86

16. Manouze H, Ghestem A, Poillerat V, Bennis M, Ba-M'hamed S, Benoliel JJ, Becker C, Bernard C (2019) Effects of single cage housing on stress, cognitive, and seizure parameters in the rat and mouse pilocarpine models of epilepsy. eNeuro 6 (4) ENEURO.0179-18.2019:1–23

17. Löscher W, Ferland RJ, Ferraro TN (2017) The relevance of inter- and intrastrain differences in mice and rats and their implications for models of seizures and epilepsy. Epilepsy Behav 73:214–235

18. Kalueff AV (2007) Mapping convulsants' binding to the GABA-A receptor chloride ionophore: a proposed model for channel binding sites. Neurochem Int 50:61–68

19. White HS (1998) Chemoconvulsants. In: Peterson SL, Albertson TE (eds) Neuropharmacology methods in epilepsy research. CRC Press, Boca Raton, pp 27–40

20. Löscher W (2011) Critical review of current animal models of seizures and epilepsy used in the discovery and development of new antiepileptic drugs. Seizure 20:359–368

21. Litchfield JT, Wilcoxon F (1949) A simplified method of evaluating dose-effect experiments. J Pharmacol Exp Ther 96:99–113

22. Löscher W (2009) Preclinical assessment of proconvulsant drug activity and its relevance for predicting adverse events in humans. Eur J Pharmacol 610:1–11

23. Kimball AW, Burnett WT, Doherty DG (1957) Chemical protection against ionizing radiation. I. Sampling methods for screening compounds in radiation protection studies with mice. Radiat Res 7(1):12

24. Mandhane SN, Aavula K, Rajamannar T (2007) Timed pentylenetetrazol infusion test: a comparative analysis with s.c.PTZ and MES models of anticonvulsant screening in mice. Seizure 16:636–644

25. Veliskova J, Velisek L (2017) Behavioral characterization and scoring of seizures in rodents. In: Pitkänen A, Buckmaster PS, Galanopoulou AS, Moshe SL (eds) Models of seizures and epilepsy, 2nd edn. Elsevier Inc., London, pp 111–124

Chapter 6

Procedures for Electrical and Chemical Kindling Models in Rats and Mice

Heidrun Potschka

Abstract

Since the first description of the kindling phenomenon, kindling paradigms have become frequently used chronic models in epilepsy research. The models are characterized by an initial induction of short focal seizures, which evolve in severity and duration with repeated exposure to the electrical or chemical trigger. In the final phase of the kindling process, animals reproducibly exhibit generalized seizures in response to stimulation. The fully kindled state is characterized by increased seizure susceptibility with a persistent lowering of seizure thresholds. While traditional kindling paradigms have been based on focal stimulation via implanted depth electrodes, alternate approaches with corneal stimulation or with repeated administration of a chemoconvulsant such as pentylenetetrazole have later been developed. Kindling models can be used to test drug candidates and to explore the pathophysiological role of a protein of interest based on kindling in genetically modified animals. Comparison of preclinical and clinical findings confirmed a high predictive validity for the testing of drug candidates in fully kindled animals. Thus, kindling models remain a mainstay of antiseizure drug development programs. Variants of kindling models with exposure to an antiseizure drug during kindling or with selection of antiseizure drug non-responders among fully kindled animals have been developed as models of drug-resistant or difficult-to-treat epilepsy. In addition, drug candidates have been administered during kindling acquisition in order to test for preventive effects. Kindling in genetically modified mice is a highly valuable approach as it can provide information about the impact of a protein on seizure thresholds, seizure initiation, spread and termination, and the generation of a hyperexcitable network.

Key words Electrical kindling, PTZ kindling, Corneal kindling, Amygdala kindling

1 Introduction: Rodent Kindling Paradigms in Experimental Epileptology

The kindling phenomenon has been initially described and named by Goddard and colleagues [1, 2]. Since its first description and characterization, it became a popular model of temporal lobe epilepsy. Electrical kindling is based on focal stimulation of susceptible brain regions. In response to first stimulations, animals exhibit a minor behavioral response associated with a short and regionally limited electrographic afterdischarge. In this phase, the kindling

Divya Vohora (ed.), *Experimental and Translational Methods to Screen Drugs Effective Against Seizures and Epilepsy*, Neuromethods, vol. 167, https://doi.org/10.1007/978-1-0716-1254-5_6,
© Springer Science+Business Media, LLC, part of Springer Nature 2021

paradigm models focal partial seizures [3]. With repeated stimulations, the response evolves with electrographic and behavioral seizures progressing in severity and duration. Thus, in later phases of the kindling process, animals exhibit seizures with focal onset and secondary generalization [3]. The progressive increase in the seizure response is the main characteristic defining the kindling phenomenon. The fully kindled state is characterized by a persistent increase in seizure susceptibility with a lowered seizure threshold, so that stimulus currents that initially only trigger short focal seizures reproducibly induce generalized tonic-clonic seizures in fully kindled animals. Despite the evident increase in excitability and the lowering of seizure thresholds, spontaneous seizures only occur following prolonged stimulation procedures lasting several weeks to months [3].

The face and predictive validity of the kindling paradigm has been analyzed in multiple studies. While kindling does not result in a pronounced neuronal cell loss, a reorganization of neuronal circuitries occurs as a consequence of repeated kindling stimulations [3–6]. In addition, astrocytic hypertrophy has been reported in kindling paradigms [3, 7]. Cellular alterations are reflected by a modulation of molecular expression patterns. Kindling-associated changes in the transcriptome proved to be complex and influenced by the antiseizure drug levetiracetam [8, 9].

As stated by Löscher [10], the responsiveness of kindled seizures to antiseizure drugs seems to predict the clinical efficacy in patients with partial seizures. Interestingly, the comparison between pharmacological data from kindled animals with those from animals with spontaneous seizures in post-status epilepticus models revealed a comparable pharmacology [10]. This, however, only applied when anticonvulsant effects of drugs are assessed in fully kindled animals with reproducible induction of generalized seizures [10]. In apparent contrast, the effects of drugs on kindling progression can largely differ from those on development of epilepsy with manifestation of spontaneous recurrent seizures in post-status epilepticus models [10].

While traditional kindling paradigms with focal electrical stimulation via implanted depth electrodes have been based on once-daily stimulations, rapid kindling procedures have later been developed in order to speed up the kindling process and allow higher throughput. Respective paradigms have applied several stimulations in short intervals within 1 day [3]. As pointed out by McIntyre [3], the success of respective approaches depends on the stimulated brain region. Whereas longer interstimulus intervals are necessary for the amygdala, rapid kindling has been successfully established with stimulation of the hippocampus [11].

The fact that electrical kindling with stimulation via implanted depth electrodes is associated with an invasive surgical procedure, efforts have been made to develop a non-invasive variant. Based on an initial description by Sangdee and colleagues [12], Matagne and Klitgaard [13] intensely characterized the corneal kindling paradigm with twice-daily stimulations via corneal electrodes.

As the technical equipment necessary for electrical kindling can constitute another hurdle, chemical kindling procedures have also been developed. The most common chemical kindling paradigm is based on repeated injections of the chemoconvulsant pentylenetetrazole (PTZ).

While corneal kindling and PTZ kindling offer advantages regarding ease of use and feasibility, the fact that these procedures are most of the time applied without electrographic recordings is unfortunate considering that parallel analysis of behavioral and electrographic seizure activity guarantees a higher level of accurateness and sensitivity.

2 Materials

2.1 Kindling via Implanted Depth Electrodes

In addition to material necessary for stereotactic surgery including a stereotaxic frame for laboratory rodents, surgical instruments, sterile towels, sterile swabs, ophthalmic ointment, 70% ethanol, 3% H_2O_2, anesthetic, analgesic, and antibiotic drugs, the following equipment are required:

- Bipolar electrodes (e.g., self-constructed from stainless steel wire) (Fig. 1a).

- Screws (e.g., three per animal for grounding electrode and to fix the electrode assembly to the skull) (Fig. 1).

- Stimulator (e.g., 2× HSE Type 215E12 from Hugo Sachs Elektronik GmbH, Hugstetten, Germany).

- Switch box (e.g., self-constructed to switch between stimulation and recording mode).

- Amplifier (e.g., BioAmp, ADInstruments Ltd., Hastings, UK).

- Analog-to-digital converter (e.g., PowerLab 4/30, ADInstruments Ltd., Hastings, UK).

- Transparent chamber or cylinder (for stimulation and observation following kindling stimulation).

- Computer including software for EEG analysis (e.g., LabChart 7, ADInstruments Ltd., Hastings, UK).

- Camera connected to computer (recommended for video recording).

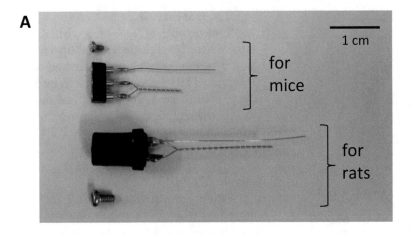

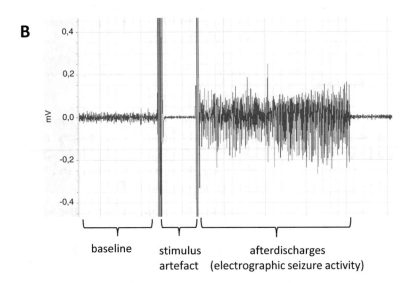

Fig. 1 (**a**) Electrode and screws, which can be used for amygdala kindling in mice and rats. (**b**) Electrographic recording from a kindled mouse with induction of a generalized seizure: baseline EEG, stimulus artifact, and afterdischarges

2.2 Corneal Kindling   In addition to the local anesthetic, the following equipment are required:

- Stimulator (e.g., BMT Medizintechnik, GmbH, Berlin, Germany, or Harvard Apparatus, Holliston, Massachusetts, USA).

- Copper electrodes for corneal stimulation (e.g., from Harvard Apparatus, Holliston, Massachusetts, USA).

- Transparent chamber or cylinder (for observation following kindling stimulation).

- Camera and computer (recommended for video recording).

2.3 Chemical Kindling

The following material and equipment are required:

- Pentylenetetrazole (PTZ) (e.g., from Sigma-Aldrich, Darmstadt, Germany).
- 0.9% NaCl.
- Transparent chamber or cylinder (for observation following PTZ administration).
- Camera and computer (recommended for video recording).

3 Methods

3.1 Kindling via Implanted Depth Electrodes

The classic electrical kindling paradigm in rats and mice is based on implantation of depth stimulation and recording electrodes.

1. Following habituation to the animal facility, the laboratory environment, and the experimenters, animals can be prepared for surgery to implant the electrode.

2. Following shaving, the implantation site is disinfected with 70% ethanol.

3. To avoid hypothermia during surgery, animals should be placed on heating mats ideally regulated by the animal's body temperature.

4. General anesthesia, e.g., based on administration of the inhalation anesthetic isoflurane, is required for the implantation procedure.

5. Additional perioperative analgesia should be applied based on non-steroidal anti-inflammatory drugs such as meloxicam or carprofen and/or opioids such as buprenorphine.

6. Skin infiltration with the local anesthetic along the surgical incision line and on the periost before scraping away the periost is recommended to limit activation of the nociceptive system, thereby contributing to a possible preventive effect.

7. Traditional electrode target regions include the amygdala and the hippocampus. Precise placement of the depth electrodes in the target brain region is guaranteed by stereotactic surgery using a stereotaxic instrument specifically developed to fix the skull of mice and/or rats. The tooth bar needs to be adjusted in order to guarantee a horizontal positioning of the skull parallel to the base of the stereotaxic instrument, so that bregma and lambda are at the same level. Exact placement of the animal in the head holder is crucial for the accurateness of the surgical procedure. Therefore, the ear bars need to be inserted in the external auditory meatus.

8. A midline incision is made starting posterior of the eyes and ending at the ears. Clips are used to hold away the skin from the skull. The skull is cleaned to expose bregma and lambda. The skull is prepared by wiping with 5% H_2O_2, which also helps to better visualize the cranial sutures.

9. Following surgical exposure of the skull, the position for the electrode hole is calculated in relation to bregma. The exact localization for the electrode tip depends on the rat strain and sex used. It should therefore be controlled and if necessary adjusted based on test surgeries with histological control of the electrode position.

10. As an orientation, one can consider the following coordinates: kindling of the basolateral amygdala in female Sprague Dawley rats, AP −2.2, LAT +4.7, and DV 8.5, or in female Wistar rats, AP −2.2, LAT +4.8, and DV 8.5, and kindling of the amygdala in C57BL/6 mice, AP −1.4, LAT +3.3, and DV 4.8.

11. The position of the hole for the electrode can be labelled with the help of a needle with the tip freshly covered with an Edding pen. A small hole for the electrode hole is then carefully drilled with a drill adjusted to the species, size of the skull, and size of the electrode.

12. Additional screws and dental cement are used to fix the electrode assembly at the skull. The grounding electrode is connected to a screw placed above the contralateral hemisphere.

13. For the stimulation procedure, the animal is placed in a transparent cylinder or chamber allowing observation of the animal's behavioral and motor response.

14. Prior to daily kindling stimulations, the initial afterdischarge threshold is determined by a staircase stimulation procedure, e.g., starting with 8μA with subsequent stepwise 20% increases in the stimulation current (1 ms monophasic square wave pulses, 50 Hz for 1 s). The afterdischarge threshold is defined as the stimulation current, which resulted in high-amplitude and high-frequency spiking of at least 5 s duration (Fig. 1b).

15. From the next day on, daily suprathreshold stimulations are performed with a pre-defined stimulation strength (e.g., 500μA or higher depending on the initial afterdischarge thresholds). As an alternate approach, individualized suprathreshold stimulations are possible, e.g., at 20% above the individual threshold.

16. For each stimulation, the following parameters are recorded: seizure severity, seizure duration, and duration of afterdischarges (high-frequency spiking with at least twice the baseline amplitude and at least 5 s duration).

17. Seizure severity is scored according to a modified Racine score [14]: 1, immobility and mild facial clonus (e.g., closing of an eye); 2, facial clonus with chewing and head nodding; 3, unilateral forelimb clonus; 4, bilateral forelimb clonus and rearing; and 5, bilateral forelimb clonus, rearing, and falling. Stimulations should always be performed at the same time of the day.

18. Video recordings are highly recommended as these allow analysis of behavioral seizures by an additional independent experimenter.

19. Stimulations are continued until ten consecutive stage 5 seizures are observed. Considering that mice do not reproducibly show stage 5 seizures, mice are kindled until they reach ten consecutive stage 4 or 5 seizures.

20. Depending on the research question, the kindling threshold can be determined again at the end of the kindling procedure to assess the post-kindling threshold. Typically, the animals should exhibit lowered thresholds reflecting the increase in seizure susceptibility with ongoing stimulations. Fully kindled animals can then be used repeatedly for drug testing procedures as further explained below.

3.1.1 Notes

1. The equipment for electrical kindling comprises stimulators (e.g., 2× HSE Type 215E12 from Sachs, Hugstetten, Germany), a switch box, an amplifier (e.g., BioAmp, ADInstruments Ltd., Hastings, UK), and an analog-to-digital converter (PowerLab 4/30, ADInstruments Ltd., Hastings, UK). The current strength at different stimulation currents should be checked on a regular basis, and a calibration should be completed before each project.

2. Please note that very gentle procedures are necessary in mice for the insertion of electrodes in order to avoid injury. It is highly recommended that an instruction and demonstration by experienced experimenters is planned for persons in training.

3. Aseptic conditions should be implemented throughout the procedure and for all material used during surgery, including implanted electrodes and screws, to limit the risk of infection during surgery. In addition, perisurgical antibiotic treatment may be considered depending on the sensitivity of the species and strain to pathogens. The choice of the antibiotic needs to consider the tolerability in rats and mice taking into account that these species tend to develop intestinal dysbiosis in response to drugs affecting gram-positive microbiota.

4. As discussed above, perioperative analgesia should be applied such as meloxicam or carprofen and/or opioids such as buprenorphine. Recently, it has been reported that buprenorphine tended to be superior to the non-steroidal anti-inflammatory

drugs meloxicam and carprofen for pain management in mice subjected to craniotomy [15]. However, the study left several questions open including the need for optimization of doses and administration intervals as well as the combination of compounds in a multimodal analgesic regime.

5. In addition to the administration of analgesic drugs, additional local anesthesia is recommended based on the author's experience. Thereby, the local anesthetic is applied by infiltration of the suture line and administration to the periost following skull exposure.

6. Timing of the administration of analgesics and local anesthetics before surgery needs to consider recommended pretreatment times, which consider the time interval between administration and onset of action.

7. Post-operative management should avoid therapeutic gaps in the analgesic management as these can trigger pain sensitization processes.

8. For the planning of the anesthetic and analgesic management and the selection of drugs, doses, and administration intervals, the reader is referred to respective recommendations [16–18].

9. Care should be taken to administer an eye ointment before surgery to protect from dry eye damage or discomfort.

10. As an alternate to amygdala kindling, kindling via electrodes implanted in other brain regions such as the perirhinal cortex, the piriform cortex, and the hippocampus has been frequently used. When selecting a brain region, it needs to be considered that kindling rates and the stability of the kindled state and post-kindling thresholds may vary depending on the brain region [3].

11. A recovery phase of at least 2 weeks is allowed before starting the kindling paradigm. In this context, it needs to be considered that the necessary recovery time may depend on the invasiveness of the surgical procedure and on post-surgical care. Thus, it should be determined in a laboratory-specific manner. In a recent study, we observed hyperactivity in electrode-implanted rats 2 weeks following surgery [19]. This result may indicate that a reconvalescence phase of more than 2 weeks is required before starting the kindling process.

3.2 Corneal Kindling

Corneal kindling has been established in mice. In view of animal welfare considerations, the limited stability of the kindled stage in corneal kindled animals, and the fact that one cannot stimulate in short intervals in order to determine a seizure threshold, one should critically question the need of this model. An animal welfare-based model prioritization is highly recommended in this context. Recent data from the amygdala kindling paradigm indicate that the distress associated with electrical kindling via depth elec-trodes is rather limited at least in rats [20]. Therefore, one should consider respective approaches as an alternate to corneal kindling. From the author's point of view, corneal kindling should rather be used as an exception. It may, for instance, be the method of choice in genetically modified animals with alterations in development and volume of brain regions, which do not allow a direct comparison to controls related to the impossibility to place electrodes in a compa-rable manner in wild-type and genetically modified mice. It is emphasized that a higher workload related to implantations in traditional kindling paradigms with stimulation via depth electro-des cannot be a reason to apply corneal kindling. However, it also needs to be emphasized that the severity of the corneal kindling paradigm has not been thoroughly assessed yet. In the first descrip-tion of the paradigm, the authors reported no microscopically detectable lesions of the cornea [13]. In a small number of animals, retinal lesions were observed. However, the authors pointed out that respective lesions also occur in mice without corneal kindling, so that there was no evidence for a functional link.

1. Following habituation to the animal facility, the laboratory environment, and the experimenters, corneal kindling stimula-tions can be initiated.

2. Eyes should be anesthetized with a local anesthetic before stimulations.

3. Mice are stimulated twice daily with an inter-stimulation inter-val on 12 consecutive days. The stimuli (3 mA, 3 s, pulse frequency 50 Hz) are applied via corneally placed saline-soaked copper electrodes (stimulator: BMT Medizintechnik, Berlin, Germany).

4. Immediately following stimulation, the animal is placed in a transparent observation cylinder or chamber.

5. Seizure severity is scored according to a modified Racine scale as described above. An additional score of 6 has been intro-duced for corneal kindling paradigms indicating the occurrence of tonic hindlimb extension. Motor seizure duration should be documented for all stimulations. Video recordings are highly recommended as these allow analysis of behavioral seizures by an additional independent experimenter.

6. Fully kindled mice can be used for drug testing procedures as further explained below.

3.2.1 Notes

1. The stimulations should always be applied at the same time of the day (e.g., between 8–10 am and 2–4 pm) to limit circadian influences.

2. Care should be taken to select a local anesthetic drug and formula that does not cause irritation to the eye.

3.3 Chemical Kindling

Chemical kindling has been developed based on acute chemical seizure test, which has informed about the proconvulsant effects of various compounds. The kindling procedure is based on repeated administration of an initially sub-threshold concentration of a chemoconvulsant. Among the chemoconvulsants used for kindling procedures, pentylenetetrazole is standing out based on its most common use [21]. Pentylenetetrazole (PTZ) acts as an antagonist at GABA$_A$ receptors. Interpretation of data from the PTZ kindling paradigm needs to consider the exposure to the GABA$_A$ receptor antagonist as a possible confounding factor.

1. Following habituation to the animal facility, the laboratory environment, and the experimenters, PTZ kindling can be initiated.

2. The body weight of all animals is repeatedly determined on all experimental days.

3. The administration of the PTZ dose is based on the body weight determined on the same day. Recommended injection volumes are 3 mL/kg for rats and 10 mL/kg for mice.

4. The kindling procedure is based on intraperitoneal administration of PTZ in the left or right anterior quadrant of the abdomen avoiding injections close to the midline.

5. Following PTZ injections, animals should be placed in a transparent observation cylinder or chamber. Following every stimulation, animals should be observed for at least 30 min. Decisions about the observation time following injection need to consider that prolonged or repeated seizure activity may occur following PTZ administration. Ideally, the animal behavior should be assessed for 24 h as recommended by Shimada and Yamagata [22].

6. As the behavioral and motor patterns of PTZ-induced seizures differ from those observed with electrical kindling, specific seizure scoring systems have been developed. Shimada and Yamagata [22] recently recommended the following scoring system for PTZ kindling in mice: 0, normal behavior; 1, immobilization and lying on the abdomen; 2, head nodding and facial, forelimb, or hindlimb myoclonus; 3, continuous whole-body myoclonus, myoclonic jerks, and tail held up stiffly; 4, rearing, tonic seizure, and falling down on its side; 5, tonic-clonic seizure, falling down on its back, wild running, and jumping; and 6, death. In this context, it is emphasized that with careful dose adjustment, mortality is not frequently associated with PTZ kindling. Alternate more detailed scoring systems have been described, which also consider the duration of seizure activity during the observation period (e.g., *see* [21]).

Video recordings are highly recommended as these allow analysis of behavioral seizures by an additional independent experimenter.

7. Fully kindled animals can be used repeatedly for drug testing procedures as further explained below.

3.3.1 Notes

1. Care should be taken to vary the injection site with repeated injections considering that it cannot always be avoided that small hematomas develop, which can render repeated injections increasingly painful for the animals. In severity assessment sheets, one should therefore consider to control the abdominal tone on a regular basis.

2. During kindling procedures, PTZ is often applied intraperitoneally in 24, 48, or 72 h intervals. Please note that one has to carefully check the responsiveness for a rat or mouse strain and line used in the experiment prior to the main study. It needs to be considered that even within a strain, genetic differences can exist due to genetic drift, which can significantly affect the responsiveness to chemoconvulsants. Moreover, age of the animals significantly affects the susceptibility to chemoconvulsant-induced seizures [23]. Thus, the appropriate dosing and the administration intervals have to be determined in a pilot experiment.

3. One needs to make sure that the dose is an initially sub-threshold dose in the animals used and that repeated administration results in the induction of seizures with severity and duration evolving over time. Moreover, one needs to confirm that the kindling process results in a persistent increase in seizure susceptibility and lowered thresholds. Doses commonly administered range between 25 and 45 mg/kg [21]. It has been reported that higher doses result in an acceleration of kindling. However, high doses also imply the risk of an early occurrence of clonic seizures.

4. The number of injections necessary to induced generalized seizures differs depending on factors including species, strain, and age. Shimada and Yamagata have described score 5 seizures following ten injections of 35 mg/kg PTZ in C57BL/6 mice [22].

3.4 Study Designs Based on Kindling Procedures

Kindling paradigms have been frequently used to test drug candidates. In the majority of studies, compounds have been administered to fully kindled animals, which reproducibly exhibit generalized seizures in response to stimulation. In these animals, thresholds remain fairly stable, so that the response to a new anti-seizure drug candidate can be assessed. It should be considered that the stability of thresholds largely depends on the type of the kindling paradigm, with kindling via depth electrodes generally resulting in more stable kindling states, as well as the species and strain. In the author's laboratory, thresholds and the kindled state proved to be more stable in rats than in mice.

Kindling paradigms have been frequently used for assessment of novel drug candidates. In fully kindled animals, thresholds can be repeatedly determined allowing to compare between trials with vehicle and drug administration. Provided that thresholds are stable, one can, for instance, administer vehicle and then determine thresholds with the procedure described above, i.e., repeated stimulations applied in 1 min intervals with a stimulation current increasing by 20% per stimulation. Based on this procedure, one can determine the threshold for the induction of afterdischarges (Fig. 2a). In addition, if the respective threshold stimulation is only associated with a focal seizure, stimulations can be continued until a generalized seizure is induced to determine the generalized seizure threshold. Two or 3 days later, a test compound can be administered prior to threshold determination in order to determine the anticonvulsant effect based on an impact on thresholds and on seizure parameters at threshold stimulation (Fig. 2a–c). The assessment of thresholds following vehicle or drug administration can be repeated so that it is feasible to determine a dose dependency, the dose-response curve, and the effective dose 50, which increases the threshold by 50%. Analysis of the impact on thresholds and seizure parameters is of particular interest as it provides information about ictogenesis including seizure generation and spread. While an effect on thresholds confirms an impact of a drug candidate on seizure initiation, assessment of seizure severity and duration provides information about an influence on seizure spread and seizure termination.

Thus, threshold determination following vehicle and drug administration should be considered superior and more informative in comparison with study designs limited to the assessment of seizure parameters following suprathreshold stimulation. In this context, it needs to be considered that repeated threshold determination is only possible in animals kindled via an implanted depth electrode.

In this context, it is interesting to note that one can also assess proconvulsant effects in the kindling paradigm based on a lowering of thresholds [24]. Moreover, it has been reported that fully kindled animals can be more susceptible to develop spontaneous seizures in response to administration of proconvulsant compounds [25].

Repeated exposure to a standard antiseizure drug can allow selection of responders or non-responders, which do or do not exhibit threshold increases in response to drug administration. Selection procedures have been based on three to four consecutive kindling sessions with repeated comparison of the response to phenytoin versus vehicle [26]. The group of kindled phenytoin

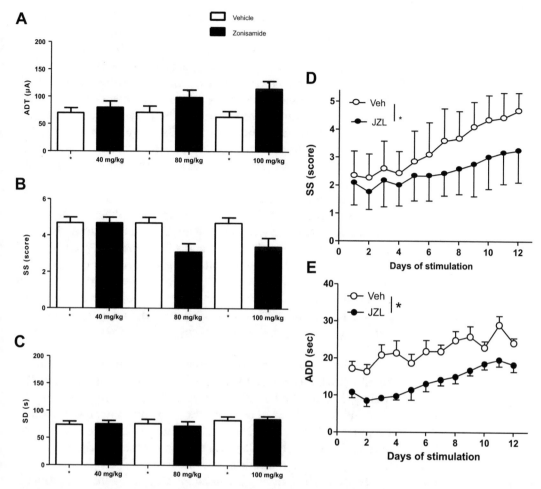

Fig. 2 Example: testing of drugs or drug candidates in the kindling paradigm. (**a–c**) Impact of perampanel on thresholds and on seizure parameters at threshold stimulation in fully kindled rats. (**d**, **e**) Impact of the monoacylglycerol lipase inhibitor JZL184 on amygdala kindling progression in mice with effects on development of seizure severity and afterdischarge duration
ADT Afterdischarge threshold, *SS* seizure severity, *SD* seizure duration, *ADD* afterdischarge duration. Data are illustrated as mean ± SEM with asterisks indicating significant differences ($p < 0.05$)
Data for perampanel from [34] with permission by John Wiley and Sons via Copyright Clearance Center. Data for JZL184 from [31] with permission by Elsevier via Copyright Clearance Center

non-responders has been suggested as a model of drug-resistant epilepsy allowing to assess the efficacy of novel drug candidates against difficult-to-treat seizures. When applying this approach, it is of utmost relevance to confirm that plasma concentrations of phenytoin are in the therapeutic range.

A respective selection procedure has also been tried in corneal kindled mice. However, the attempt failed regarding the selection of a relevant number of animals, which could be categorized as non-responders [27].

Drug candidates can also be applied during the kindling acquisition phase. The aim of respective studies is to determine possible preventive and/or disease-modifying effects. The analysis is based on a comparison of the progression of seizure severity and duration of behavioral and electrographic seizures between vehicle- and drug-treated animals (Fig. 2d, e). It is recommended to compare the number of stimulations to reach the different seizure severity scores as well as the cumulative afterdischarge duration until first occurrence of the different seizure severity scores. It needs to be considered that test compounds can affect the response to kindling stimulations and thereby delay the occurrence of generalized seizures based on an anticonvulsant effect. Thus, a re-test with additional stimulations following a withdrawal phase is necessary. Animals are then tested days following the last drug administration in order to assess whether persistent effects are observed. Only in case of effects that persist despite drug withdrawal, one can conclude about potential preventive effects. However, it needs to be considered that final conclusions about possible preventive effects require testing in models with development of spontaneous seizures, e.g., following a status epilepticus or another type of brain insult.

As further discussed in Chapter 10, exposure to drugs during kindling acquisition can result in tolerance of drug resistance. This finding has resulted in the implementation of the lamotrigine-resistant kindled rat model in the Epilepsy Therapy Screening Program (ETSP) of the National Institute of Neurological Disorders and Stroke (NINDS) and in the recent description of the lamotrigine-resistant kindled mouse model [28, 29].

In both, the lamotrigine-resistant amygdala kindling model in rats and the lamotrigine-resistant corneal kindled mice, cross-tolerance or cross-resistance was observed with different antiseizure drugs [29, 30].

In addition to pharmacological approaches, genetic approaches are of interest to study the pathophysiological mechanisms of ictogenesis and of the development of a hyperexcitable network. Moreover, genetic approaches can be applied to assess a target candidate. The functional consequences of a protein of interest can be studied based on kindling in genetically modified mice with overexpression or genetic deficiency of this protein. Kindling in genetically modified animals can provide information about the role of the respective protein in seizure initiation, generation, spread, and termination, as well as in the generation of a hyperexcitable kindled network. Moreover, the combination of pharmacological and genetic approaches can provide information about the role of a target candidate for the efficacy of a drug or drug candidate. When a compound exerts an effect in wild-type but not in knockout mice, the findings point to a relevant contribution of the deficient protein to efficacy. Respective approaches have, for instance, been applied to study the role of different components of the endocannabinoid system [31, 32] or of heat shock protein 70 [33].

4   Conclusions

Kindling paradigms have been among the most common chronic models applied in epilepsy research. Different technical procedures have been developed based on repeated stimulations via depth electrodes, corneal stimulations, or injections of chemoconvulsants such as PTZ.

Kindling is characterized by a high predictive validity with a reliable prediction of efficacy against partial (focal) seizures. Thus, kindling procedures including variants with a poor drug responsiveness remain a mainstay in the preclinical screening of antiseizure drugs. Moreover, the control of seizure occurrence based on induced seizures allows the application of kindling paradigms in the characterization of genetically modified mice, providing information about the pathophysiological role of a protein or its suitability as a target candidate.

Acknowledgments

Recent research involving kindling models has been supported by Deutsche Forschungsgemeinschaft DFG PO681/8-1 and 9-1.

References

1. Goddard GV (1967) Development of epileptic seizures through brain stimulation at low intensity. Nature 214(5092):1020–1021. https://doi.org/10.1038/2141020a0

2. Goddard GV, McIntyre DC, Leech CK (1969) A permanent change in brain function resulting from daily electrical stimulation. Exp Neurol 25(3):295–330. https://doi.org/10.1016/0014-4886(69)90128-9

3. McIntyre DC (2006) The kindling phenomenon. In: Pitkänen A, Schwartzkroin PA, Moshe SL (eds) Models of seizures and epilepsy. Elsevier Academic Press, Burlington, pp 351–361

4. Parent JM, Janumpalli S, McNamara JO, Lowenstein DH (1998) Increased dentate granule cell neurogenesis following amygdala kindling in the adult rat. Neurosci Lett 247(1):9–12. https://doi.org/10.1016/s0304-3940(98)00269-9

5. Sutula T, He XX, Cavazos J, Scott G (1988) Synaptic reorganization in the hippocampus induced by abnormal functional activity. Science 239(4844):1147–1150. https://doi.org/10.1126/science.2449733

6. Represa A, Ben-Ari Y (1992) Kindling is associated with the formation of novel mossy fibre synapses in the CA3 region. Exp Brain Res 92 (1):69–78. https://doi.org/10.1007/bf00230384

7. Khurgel M, Ivy GO (1996) Astrocytes in kindling: relevance to epileptogenesis. Epilepsy Res 26(1):163–175. https://doi.org/10.1016/s0920-1211(96)00051-4

8. Potschka H, Krupp E, Ebert U, Gumbel C, Leichtlein C, Lorch B, Pickert A, Kramps S, Young K, Grune U, Keller A, Welschof M, Vogt G, Xiao B, Worley PF, Loscher W, Hiemisch H (2002) Kindling-induced

overexpression of Homer 1A and its functional implications for epileptogenesis. Eur J Neurosci 16(11):2157–2165. https://doi.org/10.1046/j.1460-9568.2002.02265.x

9. Gu J, Lynch BA, Anderson D, Klitgaard H, Lu S, Elashoff M, Ebert U, Potschka H, Loscher W (2004) The antiepileptic drug levetiracetam selectively modifies kindling-induced alterations in gene expression in the temporal lobe of rats. Eur J Neurosci 19(2):334–345. https://doi.org/10.1111/j.0953-816x.2003.03106.x

10. Löscher W (2002) Animal models of epilepsy for the development of antiepileptogenic and disease-modifying drugs. A comparison of the pharmacology of kindling and post-status epilepticus models of temporal lobe epilepsy. Epilepsy Res 50(1–2):105–123. https://doi.org/10.1016/s0920-1211(02)00073-6

11. Lothman EW, Williamson JM (1994) Closely spaced recurrent hippocampal seizures elicit two types of heightened epileptogenesis: a rapidly developing, transient kindling and a slowly developing, enduring kindling. Brain Res 649 (1–2):71–84.   https://doi.org/10.1016/0006-8993(94)91050-2

12. Sangdee P, Turkanis SA, Karler R (1982) Kindling-like effect induced by repeated corneal electroshock in mice. Epilepsia 23 (5):471–479.   https://doi.org/10.1111/j.1528-1157.1982.tb05435.x

13. Matagne A, Klitgaard H (1998) Validation of corneally kindled mice: a sensitive screening model for partial epilepsy in man. Epilepsy Res 31(1):59–71. https://doi.org/10.1016/s0920-1211(98)00016-3

14. Racine RJ (1972) Modification of seizure activity by electrical stimulation. II. Motor seizure. Electroencephalogr Clin Neurophysiol 32 (3):281–294.   https://doi.org/10.1016/0013-4694(72)90177-0

15. Cho C, Michailidis V, Lecker I, Collymore C, Hanwell D, Loka M, Danesh M, Pham C, Urban P, Bonin RP, Martin LJ (2019) Evaluating analgesic efficacy and administration route following craniotomy in mice using the grimace scale. Sci Rep 9(1):359. https://doi.org/10.1038/s41598-018-36897-w

16. Flecknell P (2018) Rodent analgesia: assessment and therapeutics. Vet J 232:70–77. https://doi.org/10.1016/j.tvjl.2017.12.017

17. Gargiulo S, Greco A, Gramanzini M, Esposito S, Affuso A, Brunetti A, Vesce G (2012) Mice anesthesia, analgesia, and care, part I: anesthetic considerations in preclinical research. ILAR J 53(1):E55–E69. https://doi.org/10.1093/ilar.53.1.55

18. Foley PL, Kendall LV, Turner PV (2019) Clinical management of pain in rodents. Comp Med 69(6):468–489. https://doi.org/10.30802/AALAS-CM-19-000048

19. Möller C, van Dijk RM, Wolf F, Keck M, Schonhoff K, Bierling V, Potschka H (2019) Impact of repeated kindled seizures on heart rate rhythms, heart rate variability, and locomotor activity in rats. Epilepsy Behav 92:36–44. https://doi.org/10.1016/j.yebeh.2018.11.034

20. Möller C, Wolf F, van Dijk RM, Di Liberto V, Russmann V, Keck M, Palme R, Hellweg R, Gass P, Otzdorff C, Potschka H (2018) Toward evidence-based severity assessment in rat models with repeated seizures: I. Electrical kindling. Epilepsia 59(4):765–777. https://doi.org/10.1111/epi.14028

21. Gilbert ME, Goodman JH (2006) Chemical kindling. In: Pitkänen A, Schwartzkroin PA, Moshe SL (eds) Models of seizures and epilepsy. Elsevier Academic Press, Burlington, pp 379–393

22. Shimada T, Yamagata K (2018) Pentylenetetrazole-induced kindling mouse model. J Vis Exp (136). https://doi.org/10.3791/56573

23. Nokubo M, Kitani K, Ohta M, Kanai S, Sato Y, Masuda Y (1986) Age-dependent increase in the threshold for pentylenetetrazole induced maximal seizure in mice. Life Sci 38 (22):1999–2007. https://doi.org/10.1016/0024-3205(86)90147-5

24. Löscher W (2009) Preclinical assessment of proconvulsant drug activity and its relevance for predicting adverse events in humans. Eur J Pharmacol 610(1–3):1–11. https://doi.org/10.1016/j.ejphar.2009.03.025

25. Potschka H, Friderichs E, Loscher W (2000) Anticonvulsant and proconvulsant effects of tramadol, its enantiomers and its M1 metabolite in the rat kindling model of epilepsy. Br J Pharmacol 131(2):203–212. https://doi.org/10.1038/sj.bjp.0703562

26. Löscher W (2016) Fit for purpose application of currently existing animal models in the discovery of novel epilepsy therapies. Epilepsy Res 126:157–184.   https://doi.org/10.1016/j.eplepsyres.2016.05.016

27. Potschka H, Loscher W (1999) Corneal kindling in mice: behavioral and pharmacological differences to conventional kindling. Epilepsy Res 37(2):109–120. https://doi.org/10.1016/s0920-1211(99)00062-5

28. Wilcox KS, West PJ, Metcalf CS (2020) The current approach of the Epilepsy Therapy Screening Program contract site for identifying

improved therapies for the treatment of pharmacoresistant seizures in epilepsy. Neuropharmacology 166:107811. https://doi.org/10.1016/j.neuropharm.2019.107811

29. Koneval Z, Knox KM, White HS, Barker-Haliski M (2018) Lamotrigine-resistant corneal-kindled mice: a model of pharmacoresistant partial epilepsy for moderate-throughput drug discovery. Epilepsia 59(6):1245–1256. https://doi.org/10.1111/epi.14190

30. Metcalf CS, Huff J, Thomson KE, Johnson K, Edwards SF, Wilcox KS (2019) Evaluation of antiseizure drug efficacy and tolerability in the rat lamotrigine-resistant amygdala kindling model. Epilepsia Open 4(3):452–463. https://doi.org/10.1002/epi4.12354

31. von Rüden EL, Bogdanovic RM, Wotjak CT, Potschka H (2015) Inhibition of monoacylglycerol lipase mediates a cannabinoid 1-receptor dependent delay of kindling progression in mice. Neurobiol Dis 77:238–245. https://doi.org/10.1016/j.nbd.2015.03.016

32. von Rüden EL, Jafari M, Bogdanovic RM, Wotjak CT, Potschka H (2015) Analysis in conditional cannabinoid 1 receptor-knockout mice reveals neuronal subpopulation-specific effects on epileptogenesis in the kindling paradigm. Neurobiol Dis 73:334–347. https://doi.org/10.1016/j.nbd.2014.08.001

33. von Rüden EL, Wolf F, Gualtieri F, Keck M, Hunt CR, Pandita TK, Potschka H (2019) Genetic and pharmacological targeting of heat shock protein 70 in the mouse amygdala-kindling model. ACS Chem Neurosci 10(3):1434–1444. https://doi.org/10.1021/acschemneuro.8b00475

34. Vera Russmann, Josephine D Salvamoser, Maruja L Rettenbeck, Takafumi Komori, Heidrun Potschka (2016) Synergism of perampanel and zonisamide in the rat amygdala kindling model of temporal loba epilepsy. Epilepsia. Apr;57(4):638–47. https://doi.org/10.1111/epi.13328. Epub 2016 Feb 8

Chapter 7

Methods for the Induction of *Status Epilepticus* and Temporal Lobe Epilepsy in Rodents: The Kainic Acid Model and the Pilocarpine Model

Shreshta Jain, Nikita Nirwan, Nidhi Bharal Agarwal, and Divya Vohora

Abstract

Status epilepticus (SE) and temporal lobe epilepsy (TLE) models are developed to understand the pathophysiology involved, discovering novel therapeutic targets and evaluating new pharmacological interventions. The in vivo animal models are generated to replicate the pathological, neurobehavioral, and electrophysiological characteristics observed in TLE patients. There are different types of models based on their method of induction of epileptic seizures. In this chapter, we describe the two commonly employed chemically induced models of SE and TLE, namely, the kainic acid and the pilocarpine model. The chapter discusses the mechanistic basis of using these chemoconvulsants, their characteristic features, and the experimental methodology involved in inducing SE through various routes of administration as employed by various researchers. The seizure scoring, electroencephalographic recording, and interpretation of neurodegenerative changes along with the troubleshooting associated with the procedure are described. Our aim is to provide guidance for the readers to induce recurrent seizures in rodents that resemble the continuous spontaneous seizures associated with TLE.

Key words Temporal lobe epilepsy, Pilocarpine, Kainic acid, Status epilepticus, EEG recording, Neurodegeneration, Hippocampus

1 Introduction

Temporal lobe epilepsy (TLE) is a type of focal epilepsy characterized by unpredictable and repeated seizures, often refractory to therapy, with origin from one of the temporal lobes of the brain. Approximately 65% of patients suffer from TLE and develop structural changes known as hippocampal sclerosis (HS). Sclerotic hippocampus is associated with neuronal loss and gliosis in the CA1 and CA3/CA4 region [1, 2]. The in vivo models for TLE are

---

Shreshta Jain and Nikita Nirwan contributed equally with all other contributors.

Divya Vohora (ed.), *Experimental and Translational Methods to Screen Drugs Effective Against Seizures and Epilepsy*, Neuromethods, vol. 167, https://doi.org/10.1007/978-1-0716-1254-5_7,

developed to resemble the histological patterns of neurodegeneration along with neurobehavioral and electroencephalographic characteristics observed in the patients diagnosed with TLE.

Epileptogenesis is triggered by *status epilepticus* (SE). According to the International League Against Epilepsy (ILAE), the SE is defined as a condition resulting either from the failure of the mechanisms responsible for seizure termination or from the initiation of mechanisms, which lead to abnormally, prolonged seizures. It is a condition, which can have long-term consequences, including neuronal death, neuronal injury, and alteration of neuronal networks, depending on the type and duration of seizures [3].

This chapter will cover about two chemoconvulsant-induced SE and TLE models, the kainic acid (KA) and the pilocarpine-induced models that involve an initial injurious stage of SE followed by a latent period following which there is emergence of repeated spontaneous recurrent seizures (SRS). These models have been broadly accepted in extensive research-based studies. The chapter initially discusses the mechanistic basis of these chemoconvulsants followed by details of methodology involved in developing SE. Post-SE induction, the procedures of behavioral seizure scoring, EEG recordings, and interpretation of associated neurodegeneration are elaborated. The required steps and troubleshootings associated with this model are also detailed in this chapter.

The characteristic features developed and common to both pilocarpine and KA model are:

(a) **Acute SE:** Identified by the occurrence of tonic-clonic seizures.

(b) **Latent period:** After several hours of the acute SE, animal enters into the latent period (seizure-free period) which represents the period occurring after an initial precipitating injury in patients with TLE [4].

(c) **Spontaneous recurrent seizures:** After a latent period of approximately 7–14 days for pilocarpine-induced model, and 5 days or up to a month, depending upon the route of administering kainic acid, spontaneous seizures are observed in both models after the first SE [5–8].

(d) **Hippocampal sclerosis and neurodegeneration:** Extensively distributed lesions and neurodegeneration are seen in the brain regions such as the hippocampus, amygdala, and piriform cortex representing mesial or cornu ammonis sclerosis seen in human TLE. After a few weeks of initial SE induction, rodent's hippocampal and parahippocampal brain region undergo the neuronal network reorganization characterized by robust mossy fiber sprouting (MFS) into the inner molecular layer of dentate gyrus (DG), interneuron loss, and ectopic dentate granule cell proliferation [4, 9, 10].

(e) **Progression of acute SE into the model of drug refractory TLE:** According to the available literature, both kainic acid and pilocarpine models imitate the attributes of the pharmacotherapeutic treatment of clinical epilepsy based on the time and duration of the intervention of antiseizure drugs. Drugs such as phenobarbital, carbamazepine, valproate, ethosuximide, lamotrigine, and levetiracetam, prevent seizures due to their anti-ictogenic potential, and their effects may vary from complete seizure reduction to no effect at all; although the plasma drug concentration remains in therapeutic range, later seizures reoccur on withdrawal of treatment, and hence certain group of animals based on their age, species, and strain become pharmacoresistant, and the seizures are not reduced even after recurrent treatment with antiseizure drugs (Fig. 1) [11–15].

1.1 The Kainic Acid Model

Kainic acid [2-carboxy-4-(1-methylethenyl)-3-pirrolidiacetic acid] is a cyclic L-glutamate analogue derived from the red algae *Digenea simplex* that was initially used as an anthelmintic to eradicate ascariasis. Later, it was discovered that kainic acid (KA) was responsible for vigorous depolarization and sustained excitatory actions in the cortical neurons of rats [16], which was analogous to the effects initiated on activation of glutamate receptors, thereby inducing apoptosis as observed in pathogenesis of TLE [17]. The kainic acid-induced SE model is a well-established model for investigating therapeutic interventions against TLE. This model is preferred in evaluating different stages of epileptogenesis, from onset of SE to the development of continuous spontaneous seizures. The neuronal structural loss known as hippocampal sclerosis (HS) is observed primarily in CA1 and CA3 regions (HS type 1) clinically seen in patients with TLE [18]. The KA model reflects relatable neuropathological changes characterized by similar neuronal loss as seen in patients diagnosed with HS type 1 [19].

Mechanistic Pathway: *Kainic acid acts as an agonist for the ionotropic glutamate AMPA receptors (non-NMDA) and at kainate-binding sites (KA receptors). The KA receptors have two subtypes—KA1 and KA2. While the levels of KA1 receptors are highly expressed in the CA3 region of the hippocampus, KA2 receptors are present in both CA1 and CA3 region of the hippocampus [7]. Hence, the pyramidal cells of CA3 region of the hippocampus are highly vulnerable to the excitotoxicity induced by KA rendering the hippocampus as the site of seizure onset in this model. The mechanism of induction of TLE implicates neuronal depolarization due to ionic influx of Na^+, K^+, or Ca^{2+} ions. The influx of monovalent ions such as Na^+ and K^+ is regulated via activation of AMPA and KA receptors. Even though NMDA receptor controls the influx of Ca^{2+} ions, previous studies stress that the excitatory action of KA is observed due to augmented Ca^{2+}*

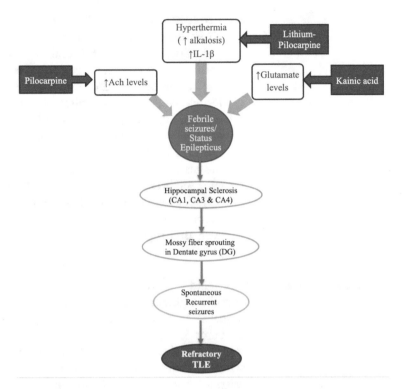

Fig. 1 Mechanistic representation of models inducing SE and TLE. Kainic acid acts on the NMDA glutamate receptors and enhances the levels of the excitatory neurotransmitter, glutamate. Pilocarpine increases the levels of acetylcholine (ACh) by activating muscarinic receptor, M1. On the other hand, lithium-pilocarpine model induces hyperthermia due to increased alkalosis, as a result of elevated interleukin, IL-1β, concentration. The administration of these chemoconvulsants induces brain injury and leads to febrile seizures with the onset of *status epilepticus* (SE). Neurodegeneration is observed post-SE in the hippocampal sub-regions of CA1, CA3, and CA4, including mossy fiber sprouting observed in dentate gyrus (DG). After a latent period, spontaneous recurrent seizures (SRS) are developed, and the animals become models for chronic, mostly refractory, temporal lobe epilepsy (TLE)

ion permeability. This can be elucidated by GluR2 subunit located in the AMPA receptors, responsible for modulating the Ca^{2+} ion influx to the neurons. Also, ablation of GluR5 subunit of KA receptors in knockout mice models have been associated with increased seizure susceptibility [20], whereas that of GluR6 subunit leads to lesser induction of seizures [21].

Model Features: *KA can be administered through different routes, namely, systemic and intracranial. Systemic administration of KA involves intraperitoneal, subcutaneous, or intravenous routes, while intracranial route of administration includes intraamygdala,* *intrahippocampal, and intracortical injection [10]. Another route of administration that has been recently explored is intranasal route [22].*

Systemic administration of KA is employed commonly for inducing SE because of its non-surgical and simple procedure. However, there are few disadvantages with systemic administration of KA over other routes of induction—(a) damage observed in extra-hippocampal parts of the brain, (b) bioavailability that cannot be controlled, (c) bilateral damage induced instead of unilateral as observed clinically, and (d) high mortality rate due to systemic side effects.

On the other hand, intracerebral method of induction by KA is considered advantageous as (a) the dose of KA can be optimized and utilized in various strains of similar species; (b) the site of induction can be controlled on the basis of the stereotaxic coordinates; (c) induction of SE is more focal in the hippocampus region, avoiding extra-hippocampal damage; and (d) the administration of KA can be controlled to evoke unilateral HS as observed in TLE patients [10, 23].

The systemic administration of KA causes mossy fiber sprouting and spontaneous repeated seizures, whereas the intracerebral administration of KA induces neurodegeneration due to excessive abnormal epileptiform discharge leading to SE in the limbic system [24, 25].

1.2   The Pilocarpine Model

Pilocarpine-induced SE has been widely used in many laboratories due to its strong parallelism with human TLE. This model was first introduced by Turski, several years ago [26]. As compared to other chemical models of developing TLE, this model is better due to its shorter latency period to develop SE (5–10 min) and its possibility to reproduce human analogous epileptogenesis in a wide range of strains in rodents [5, 27]. Pilocarpine-induced neuronal loss falls under the HS type 1 defined by ILAE (since neuronal loss is significantly seen in CA1, CA2, and CA3 regions) [28].

Mechanistic Pathway: *The mechanisms underlying pilocarpine-induced SE are known to be the activation of M1 muscarinic receptors (i.e., cholinergic system) because seizures are not developed in the M1 receptor knockout mice [29]. However, the maintenance of pilocarpine-induced seizures is governed by mechanism other than cholinergic as atropine becomes ineffective after definite time of SE induction [5]. It has been reported that these seizures are maintained by NMDA (N-methyl-D-aspartate) receptor activation, which raises Ca^{2+} in postsynaptic cells resulting in excitotoxicity and neuronal cell death [5]. Seizure progression could also be the consequences of elevated serum IL-1β (proinflammatory cytokine) that damages the blood-brain barrier [30].*

Model Features: *Pilocarpine could be administered by systemic route or intracerebral (intrahippocampal and intracerebroventricular) route. The developed behavioral, EEG, and neuropathological alterations are similar by all routes of pilocarpine administration [31]. Behavioral changes induced by systemic administration of pilocarpine are dose dependent. The dose of pilocarpine used for systemic administration could range from 100 to 400 mg/kg, and the initial characteristics experienced by rodents are immobility followed by gustatory and olfactory automatisms (salivation, orofacial movements, and vibrissae twitching).*

Systemic administration of pilocarpine is associated with the severe and rapid neurodegeneration in hippocampal as well as extra-hippocampal brain regions including CA2. The neurodegeneration beginning in the dentate gyrus (DG) hilus is accompanied by the damage in CA3 and CA1 region. The amygdala and thalamus also undergo significant damage. The neuronal loss then spread to the subiculum, septum, olfactory tubercle, piriform and entorhinal cortices, and neocortex. The rodents that develop SRS exhibit damage in areas like substantia nigra, lateral thalamic nucleus, and dentate hilus. The neuronal loss exhibited in the mossy cells of DG promotes mossy fiber sprouting [4, 5]. On the contrary, the extent of neurodegeneration induced by intrahippocampal route is regionally selective because of the activation of restricted area (hippocampal region) of brain circuitry; however, the pattern of neurodegeneration remains similar to the systemic route of administration, initially seen in the DG hilus and extending to CA1 and CA3 region. Sometimes, the minority of extra-hippocampal regions (amygdala and thalamus) also experiences neurodegeneration [32]. Animals undergoing severe SE also exhibit the neuronal loss in CA2 and the granule cell layer of DG, regions that are thought to be resistant to pilocarpine effect [32–34].

Respiratory failure following convulsions is a common cause of acute death after pilocarpine treatment [35]. Ramping up protocol is thus employed to reduce mortality and induce SE. In this method, injections of 100 mg/kg of pilocarpine are given every 10 min until onset of SE, defined as continuous seizures with at least one stage 3–5 seizures (forelimb clonus, rearing, falling), according to the Racine scale [36]. The injection is terminated if rodent does not exhibit a response by the sixth injection.

Although limited studies have been conducted on the intracerebroventricular injection of pilocarpine, this route exhibits the potential of distributing pilocarpine focally to the brain. There is a need of elaborative research for finding out the affected areas and extent of neurodegeneration developed by this route. According to previous literature, the effectiveness of intracerebral route is far better than the systemic one. Intrahippocampal and intracerebroventricular injection are capable of developing SE in 90% and 73%

of rodents as compared to the 60% by systemic route. Out of all the routes, pilocarpine administration via intrahippocampal route exhibits zero mortality and rapid recovery from SE followed by the intracerebroventricular with 35% mortality and systemic route with up to 70% mortality rates [31, 32, 37].

2 Materials

2.1 Animals

All animal experiments are conducted in accordance with the animal ethical guidelines of the institution and national guidelines as applicable. Rodents are acclimatized to the laboratory conditions before beginning of experiment and are subjected to the free access of food and water. Rats and mice are the commonly used species for developing these models. The developed features in terms of behavior, electroencephalogram (EEG), and neurodegeneration are similar in both species [5]. Due to the widely available transgenic and knockout species, mice are increasingly popular for developing these models.

Investigation performed on different strains of mice to elucidate their sensitivity to SE post-KA administration resulted with 129/SvEMS and FVB/N mice strains to exhibit excitotoxic neurodegeneration, whereas C57BL/6 and BALB/c illustrated similar patterns however, at a higher dose and confined to a particular region, mainly the hippocampus [38]. C57 strains exhibit adequate sensitivity to seizure induction at a tolerable dose of KA, even though they are resistant against neurodegeneration post-systemic administration of KA [39]; hence, C57 strain of mice is commonly used to develop KA model [40–42]. Mice strains that are highly susceptible to the pilocarpine model are C57BL/6 mice, CD-1 mice, and FVB/N mice strains [9, 35] (also see 12 in 5.3 Notes section).

Although large numbers of studies have been carried out using pilocarpine in Wistar rats, studies on the Sprague Dawley rats have been demonstrated to have low mortality rates. Induction of SE by administration of KA has been reported in Sprague Dawley and Wistar Hannover strains of rats where the mortality was observed to be higher in WH strain following the classical Hellier protocol; however, the mortality rate reduced after modifying the protocol and administering the KA in low repeated doses [2, 43]. Recent studies have also demonstrated KA-induced SE in guinea pigs [41, 44].

Gender of the rodents must be cautiously selected in studies related to epilepsy. In order to avoid the variable effect of fluctuating female hormones (testosterone and estradiol) on seizure susceptibility, latency, and survival rates, male mice and rats are mostly preferred for these models [45].

2.2 Chemicals

1. Kainic acid (monohydrate) and pilocarpine (hydrochloride) can be purchased commercially. Sterile saline solution (0.9% NaCl solution) is required to dissolve both pilocarpine and kainic acid.

2. To prevent peripheral side effects on activation of cholinergic receptors by pilocarpine, anticholinergic agents such as atropine methyl bromide or α-methyl scopolamine are also required.

3. Anesthetics such as isoflurane, lidocaine, ketamine, and xylazine along with analgesics such as buprenorphine are required during surgery so as to minimize the pain.

4. Diazepam is to be administered to inhibit the effects of convulsant and to stop seizure activity.

5. Other surgical requirements include 70% ethanol used as a disinfectant, Betadine for post-surgical healing of wounds, cotton swabs, surgical sutures, and needles for micro-lance injection, having a bent tip. Formalin (10% v/v solution of 37% formaldehyde) is used to store brain tissues for histopathological studies.

6. The stain used for preparing slices of brain histopathology can be any of the following based upon the degree of neurodegeneration to be determined:

    (a) Hematoxylin and eosin (H&E) stain.

    (b) Cresyl violet stain.

    (c) Luxol fast blue (LFB) stain.

    (d) Fluoro-Jade C stain.

    (e) Timm stain.

    (f) Bielschowsky silver stain.

2.3 Surgical Equipment

1. For the procedure involving craniotomy, the stereotaxic apparatus equipped for the animal in use is required along with a large holder for probe.

2. Surgical tools such as scalpel, micro scissors, and forceps are basic requirements.

3. A manual drill is used to drill holes in the skull for implantation of electrodes used to monitor EEG recordings and to administer convulsant via intracerebral routes. Electrodes wired or non-wired can be used to place for recording as well as reference electrode.

4. During the surgery, the animal is kept under controlled conditions in a continuous anesthesia state which can be achieved by utilizing an induction box for isoflurane administration.

5. The temperature of the animal is maintained by employing heating pads. Tuberculin syringes (1.0 mL) are required for injections and administering the convulsant.

3 Methodology

3.1 Induction of Status Epilepticus Using Kainic Acid (also see Notes section 5.1)

3.1.1 Intraperitoneal Route

1. This method of induction of SE is described by Hellier and his coworkers. It simply involves administration of KA in a saline-based solution at a dose of 10–40 mg/kg for different strains of mice or 5–10 mg/kg for rats, through intraperitoneal injections.

2. Intraperitoneal administration of KA can be employed at a low dose every half an hour (5 mg/kg for mice; 5–9 mg/kg for rats) or a single injection at a high dose (20 mg/kg for mice; 12–18 mg/kg for rats) [25, 46]. The low-dose model is used when animals suffer extreme motor activity or fatigue to lessen the possibility of toxicity or mortality.

3. The KA is injected every hour while the animal is monitored constantly until a stable SE state is observed that remains so for at least 3 h; mortality rate is observed to be reduced to 5–6% versus 20–30% when high dose is administered in a single injection [25, 46].

3.1.2 Subcutaneous Route

1. Subcutaneous route of administration involves similar doses of KA as intraperitoneal. The KA is dissolved in saline and is injected subcutaneously.

2. The process of injecting KA is continued for 8 h with consequent injections at every 1 h. The induction of SE is confirmed with minimum four consequent seizures within an hour.

3. To terminate the seizures, diazepam (10 mg/kg for mice) can be administered followed by lorazepam (0.25 mg for rats; 6 mg/kg for mice) to reduce harm to animals [47].

4. Another subcutaneous model involves administering half dose of KA (2.5 mg/kg for rats) along with lorazepam in order to reduce systemic adverse effects and increase survival of SE-induced animals [48].

5. For the development of higher stages of SE from low-grade seizures, the dose of KA is gradually increased—6–12 mg/kg for mice. For the induction of SE at a single administration, higher dose of 20–30 mg/kg is used in mice.

3.1.3 Intracerebral Administration

1. Stereotaxic apparatus is required for injecting KA through this route of administration. Intracerebral administration is subdivided into intraamygdala, intrahippocampal, or intracortical, based on the site of injections.

2. The process involves preparation of a guide cannula and injecting cannulas from polyimide-coated silica capillaries similar to those used in gas chromatography with diameter 300μm and 170 μm respectively (see point 6 of Notes section 5.3).

3. The coordinates for directing the guide cannula to the site of injection is obtained from the stereotaxic brain atlas by Paxinos and Watson [49]. These coordinates also differ among species.

4. The coordinates for rats are as follows: *intraamygdala*, 750 ng/ kg of KA into the right basolateral amygdala [anteroposterior (A), −2.3 mm; lateral (L), −4.5 mm from bregma; ventral (V), −7.2 mm from the surface of the skull] [50], and *intrahippocampal*, KA injected in a volume of 0.4 μg/0.2 μL at a rate of 0.1 μL/min [A, −5.6 mm; L, ±4.5 mm; V, −5.5 mm relative to the bregma] [51].

5. The coordinates for mice are the following: *intraamygdala*, 0.3 μg of KA in 0.2 μL of phosphate buffer solution injected in the basolateral/central amygdaloid nucleus [A, −0.94 mm; L, −2.85 mm from bregma] [52], and *intrahippocampal*, 100 nL of 20 mM KA solution in saline injected dorsally or ventrally—coordinates for dorsal intrahippocampal injection are A, −2.0 cm; L, 0.125 cm; and V, −0.16 cm from bregma, and coordinates for ventral intrahippocampal injection are A, −0.36 cm; L, 0.28 cm; and V, −0.28 cm from bregma [53].

3.1.4 Intranasal (IN) Administration

1. This model has been developed recently in mice. The animals are anesthetized by keeping them in an induction box, and 2% isoflurane gas is passed within the chamber.

2. The KA dissolved in phosphate buffer saline (PBS) (10 mg/ mL) is administered in the nostrils of the mice during the state of anesthesia.

3. A total dose of 70–90 μL is instilled in the nostrils of mice at repeated doses of 30 mg/kg of KA within an interval of 2 min.

4. The animals are left undisturbed during intranasal application by keeping them anesthetized.

5. The mice are allowed to wake up after 5 min of complete instilment of KA [22, 54].

3.2 Induction of Status Epilepticus Using Pilocarpine (see Notes section 5.1 and 5.2)

3.2.1 Intraperitoneal Route

1. Initially, the rodents are pretreated with anticholinergic such as atropine sulfate (5 mg/kg, ip) or alpha-methyl scopolamine (1–2 mg/kg, ip) before 15–30 min of pilocarpine treatment to block the peripheral cholinergic side effects of pilocarpine [5] (see point 7 of Notes section 5.3).

2. After 30 min, the calculated dose of pilocarpine hydrochloride (in range of 100–400 mg/kg) dissolved in the normal saline is administered by intraperitoneal route at a specified rate based on the weight of the rodents.

3. Immediately after the pilocarpine treatment, animals are kept under observation and video recorded for the behavioral seizure scoring for 2–3 h.

4. After 90–120 min of behavioral seizure recording, single injection of diazepam injection (5–10 mg/kg) is given to terminate the seizures. Diazepam injection is repeated if seizures remain uncontrolled (see point 8 of Notes section 5.3).

3.2.2 Intrahippocampal Route

1. In this procedure, the rodents are anesthetized and mounted on a stereotaxic apparatus for the implantation of stainless steel guide cannula in the hippocampus.

2. The coordinates for directing guide cannula in rodents are as follows: A/P, −6.0; M/L, +5.3; and D/V, 4.5 mm (bregma as reference) for rats and A/P, −1.9 mm; M/L, ±1.5 mm; and D/V, −2.3 mm (bregma as reference) for the mice.

3. After this, the rodents are allowed to recover for 5–7 days before the intrahippocampal administration of pilocarpine through the cannula.

4. The pilocarpine is infused at the rate of 2.4 mg/μL, 1μL/site in rats, and 20 μg/0.2 μL/site in mice [55, 56]. After the recording of behavioral seizure scoring, rodents are subjected to the injection of diazepam to terminate seizures.

3.2.3 Intra-cerebroventricular Route

1. The procedure begins by anesthetizing rodents and mounting them on the Stoelting stereotaxic apparatus followed by placing incisor bar at −3.3 mm and attaching injection pump to above stereotaxic frame.

2. Just above the right lateral ventricle of the brain, a bore is drilled into the rodent's skull at the following coordinates: A/P, −4.1 mm; lateral, −5.2 mm; and ventral, 7 mm in rat (relative to bregma).

3. Using the injection pump, single dose of pilocarpine is infused at the rate of 2.4 mg/2 μL, with the flow rate of 1 μL/min. This procedure is followed by the seizure scoring on Racine scale [57].

3.3 Behavioral Seizure Scoring

Seizure scoring is performed on the basis of previously designed Racine scale or modified Racine Scale [58] as under:

Stages	Description
Stage 0	Normal activity
Stage 1	Rigid posture or immobility
Stage 2	Stiffened, extended, and often arched (Straub's) tail
Stage 3	Partial body clonus, including forelimb or hindlimb clonus or head bobbing
Stage 3.5	Whole body continuous clonic seizures while retaining posture

(continued)

Stages	Description
Stage 4	Rearing
Stage 4.5	Severe whole body continuous clonic seizures while retaining posture
Stage 5	Rearing and falling
Stage 6	Tonic-clonic seizures with loss of posture or jumping

A novel probable stage was observed during the experiments performed in our laboratory suggesting a state of circling movement before the onset of whole body clonic seizures (stage 3.5).

SE is defined by continuous seizure activity for at least 2 h involving stage 3.5 seizures and one stage 5 or 6 seizure or several stage 4.5 seizures [36]. To terminate the acute seizures after 2–3 h of SE induction, rodents are subjected to single or repeated injection of diazepam (10 mg/kg, ip). After this, 1 mL of 5% dextrose solution and moist chow is given to rodents for facilitating rapid recovery (also see points 1, 5, 9, 10 and 11 in Notes section 5.3).

3.4 Surgical Implantation of Electrodes for EEG Recording

1. After 4 weeks of SE initiation, rodents are prepared for the surgical electrode implantation to perform EEG monitoring.

2. Prior to surgery, rodents are anesthetized with a mixture of ketamine and xylazine solution dissolved in saline and placed on the stereotaxic frame with ear bars (see 2 of 5.3 in Notes).

3. Implantation of the bipolar Teflon-isolated stainless steel stimulation electrode into the dentate gyrus (DG) (A/P +3.9, L +1.7mm, and V +4 mm) is performed according to the atlas of Paxinos and Watson [61].

4. One screw, placed above the left parietal cortex, served as ground electrode. Two additional screws and dental acrylic cement are used to anchor the entire implant. EEG electrodes are wired to a head mount that is also fixed to the skull using dental cement (see 3–4 of 5.3 in Notes).

5. Post-operatively, animals are given gentamycin (antibiotic) and ketoprofen (analgesics) to prevent infection and pain and are allowed to recover for 5–8 days. Animals should be monitored until they become fully ambulant.

3.5 Histopatho-logical Analysis of Neuronal Loss

1. After the completion of video-EEG monitoring, the animals are euthanized using ethically approved methods such as carbon dioxide inhalation.

2. Following euthanasia, the brains are dissected out by breaking the skull. The brains can be stored in 4% paraformaldehyde fixing solution or in a solution of 10% formalin. They are further embedded in the paraffin wax.

3. Coronal sections of the appropriate size according to the strain used (e.g., 10µm for mice brains) are sliced parallelly using a microtome through the hippocampal region at a level posterior to bregma as stated in brain atlases by Paxinos [49, 59].

4. The slides of the coronal sections are stained using different types of stains that aid in determining neuropathological changes in the hippocampal region, such as hematoxylin and eosin (H&E) stain used as the primary evaluation of neurological tissues by displaying the cellular features; 0.1% cresyl violet stain for observing the loss of Nissl substance (chromatolysis) due to axonal injury; Luxol fast blue (LFB) stain for detecting demyelination in the CNS and which can be used in combination; Fluoro-Jade C staining performed to visualize degenerating neurons; Timm staining for quantifying mossy fiber sprouting; and Bielschowsky silver staining to identify pathological deposits [60].

5. Polyionic slides are preferably used, and after staining, they are rinsed, dried out in ethanol, and then submerged in xylene.

6. The prepared stained slides are analyzed under an optical microscope for neuronal loss and pyknotic cells and nuclei.

7. The percentage of pyknosis is calculated (total number of pyknotic cells/total number of cells $\times$ 100), and pyknotic nuclei or dead cells are determined in the CA1, CA3, and DG regions of the hippocampus.

4 Observations

4.1 EEG Recording Monitoring and Analysis

EEG monitoring aims at detecting the occurrence of spontaneous recurrent seizures (SRS). EEG activity is observed and video recorded using the digital video EEG system for 2 weeks (24 h/day) [61]. EEG recordings immediately after SE show that pilocarpine can evoke both ictal and non-ictal epileptic events. Non-ictal discharges are distinguished from ictal discharges on the basis of waveform morphology, frequency, and the associated behavioral alterations. In the beginning stage, low-voltage, fast activity appears in the neocortex and amygdala, while a clear pattern of theta rhythm is seen in the hippocampus. When the behavioral manifestations become more severe, high-voltage, fast EEG activity replaces the hippocampal theta rhythm. Moreover, at later stages, animals develop electrographic seizures, characterized by high-voltage, fast activity and prominent high-voltage spiking that precedes seizures. This activity appears to originate in the hippocampus and to propagate to the amygdala and neocortex [26]. Frequency and power analyses of EEG data are performed by uploading the data to automated program for EEG analysis [18]. EEG monitoring exhibits patterns in the first half hour post-systemic administration of KA. Interictal spikes are observed in the entorhinal cortex that are incident to wet-dog shakes [62]. The ictal discharges may be of no relation to the clinical symptoms; however, these are observed in the EEG graph, to appear in the CA3 sub-region and in the amygdala, further propagating to the thalamus, CA1, and frontal cortex regions. EEG activity is also detected in the hippocampus and identified at gamma frequency ranging from 30 to 40 Hz; subsequently sporadic spikes are characterized. In conclusion, the onset zone for seizure activity is the hippocampus induced by KA [63, 64]. The hippocampus is also the region required for the proliferation if KA-induced seizures as confirmed in the epileptiform EEG recordings post-intrahippocampal and intra-amygdaloid administration of KA that exhibit a similar pattern whereby the ictal spikes originate at the site of injection, the hippocampus and amygdala, respectively, and are further transmitted to contralateral amygdala and frontal cortex [7, 44, 65, 66].

4.2 Seizure-Induced Neurodegeneration

Rodents exhibiting SE for several hours experience the histopathological alterations that are localized within the olfactory cortex, amygdala, thalamus, hippocampal formation, and neocortex. Neurodegeneration can be seen in both models in the isolated brain of rodents that are sacrificed either up to 72 h after the onset of first SE or 6 weeks later, after completing EEG monitoring. Table 1 summarizes the comparison between the features of seizures and neurodegeneration following some widely used routes of administration of pilocarpine- and kainic acid-induced TLE in rodents.

In case of KA, intracerebral administration exhibits varying degrees of neurodegeneration depending upon the site of injection. The effect of intraventricular route of administering KA is markable in the CA3 and CA4 regions of the hippocampus, keeping the CA1 and dentate gyrus unaffected [67]. Unilateral intrahippocampal administration even after injected at distal sites, such as CA1 region, exhibited primary damage across the ipsilateral side in the CA3 and CA4 regions that was propagated to the whole of the hippocampus and DG [6, 68, 69]. Intraamygdala administration of KA initiates neuronal loss at the site of injection and in the ipsilateral dorsal hippocampal sub-regions of CA3/CA4 as studies have shown that pyramidal cells of these regions are more sensitive to KA during the first 48 h [69, 70]. Further damage is observed in the contralateral hippocampus, thalamus, contralateral amygdala, and neocortex after a period of 4 days if the animal survives [7, 69, 71]. Systemic administration of KA exhibits similar yet more severe damage than the intracerebral administration of KA. The neuronal damage is observed in the pyramidal cells of the CA1 and CA3 hippocampal sub-regions. Following the occurrence of SE, neuronal damage is observed in the entorhinal cortex, subiculum, claustrum, thalamus, and cerebral cortex within a period of 24 h [72]. Further neurodegeneration is observed in the brains of animals post-48 h of

Table 1
Comparison between the widely used routes of administration of pilocarpine- and kainic acid-induced TLE in rodents

Animal model	Seizure type	SRS and MFS presence	Mortality	Neurodegeneration	Recovery time after SE	Limitation
Systemic pilocarpine	Limbic SE and tonic-clonic seizures	Yes	70%	Broad area of brain	Days	High mortality; extra-hippocampal damage
Intrahippocampal pilocarpine	Limbic SE and tonic-clonic seizures	Yes	0	Restricted to hippocampal region	Hours	Need expertise for surgery; multiple lesion development in brain
Systemic KA	Convulsive and non-convulsive seizures	Yes	5–30%	Hippocampal and extra-hippocampal regions	Days	High mortality; more extensive damage compared with other models
Intrahippocampal KA	Limbic seizure-associated behavior and generalized convulsions; non-convulsive in mice	Yes	12%	CA3 and CA4 region of hippocampus	Weeks	Variable frequency and severity of spontaneous seizures; not all neural damage comes from seizures
Intra-amygdaloid KA	Limbic and generalized tonic-clonic seizures	Yes	55%	Confined to hippocampus	Weeks	An acute monophasic disorder, which differs from chronic neurodegenerative disorders; cannot mimic the pathogenesis of neurodegeneration

expressing vigorous convulsions, in the piriform cortex, olfactory bulb, substantia nigra, and dentate gyrus [7, 24]. Hence, the extent of neurodegeneration is proportional to the propagation of seizure activity and resembles to the neuronal damage observed in HS type 1 of clinical TLE.

In case of pilocarpine-induced SE, cell death and neurodegeneration in brain hippocampal region can be seen after 48–72 h. Post-systemic administration of pilocarpine, several areas appear swollen and edematous, and many cells are dark and shrunken. In addition, injured neurons, mainly interneuron, can be found primarily in DG followed by the hippocampus (CA1 and CA3 stratum pyramidale and radiatum), amygdala, and piriform cortex, similar to the damage observed in TLE patients [5]. Severe damage is observed in the substantia nigra, lateral thalamic nucleus, and dentate hilus in animals developing SRS [4].

On the contrary, intrahippocampal administration exhibits neuronal loss confined to the site-specific region that is the hippocampus. Few instances express neurodegeneration in extra-hippocampal region such as the amygdala and thalamus [32]. However, the pattern of neuronal damage remains similar to that observed via systemic administration, initiating in the DG hilus and progressing to CA1 and CA3 regions of the hippocampus. Pilocarpine-resistant regions such as CA2 and the granule cell layer of DG are also observed to suffer damage in animals experiencing severe SE [33, 34].

5   Notes

5.1 When to Administer Therapeutic Treatment?

Both pilocarpine- and kainic acid-induced animal models are identified to be post-status epilepticus model of TLE due to the generation of SRS and chronic epilepsy and can be employed to investigate both epileptogenesis and anti-epileptogenesis. The time of initiating a treatment and its duration depends upon the "therapeutic window," that is, the latent period for the model selected. The treatment drug can be administered prior to the onset of SE, that is, before the administration of chemoconvulsants, that may result in attenuating the severity or reducing the duration of the SE and, hence, prevent long-term consequences due to a brain insult. The alternate and clinically more relevant method to investigate anti-epileptogenesis is administering the drug after the onset of SE in order to evaluate prolonged prophylactic effect [73].

5.2 How Is Lithium-Pilocarpine Model Different?

Pilocarpine combination with lithium potentiates the convulsive activity of the pilocarpine in rats, hence increasing the ability of SE induction with reduced latency to seizure generation [74]. Lithium chloride (3 mEq/kg, ip) is generally administered 24 h before the pilocarpine administration, and it allows a conspicuous reduction of the pilocarpine dose required to induce seizures (i.e., from about 300 mg/Kg to 30 mg/Kg). The proposed mechanism behind such effect of lithium is the increased serum interleukin-1β (peripheral inflammatory mediator) levels and disruption of BBB which potentiates the seizure development [75].

The features developed by lithium-pilocarpine and high-dose pilocarpine treatments are behaviorally, electrographically, and neuropathologically indistinguishable. The rate of developing tonic-clonic seizure is 100% in lithium-pilocarpine model, while it's found to be 60% in high-dose pilocarpine model [8]. EEG waves are characterized by single spikes followed by generalized spike activity separated by intermittent low-voltage activity, after which spike trains become continuous. In few animals, the ventral forebrain is the site of origin for electrographic seizures, while in others seizures begin simultaneously at multiple sites. Neither lithium (3 mEq/kg) nor pilocarpine (30 mg/kg) caused abnormal EEG responses when administrated alone. Neurodegeneration seen in this model is similar to the one observed in high-dose pilocarpine models [76]. Development of the neurodegeneration in the areas outside the hippocampus supports its role in studying epileptogenesis [77].

As mortality of animals remains high in lithium-pilocarpine model, the introduction of modified protocol (ramping up protocol) involving administration of divided doses of pilocarpine is found helpful. Post-lithium administration, if pilocarpine is given in divided doses of 10 or 5 mg/kg at 30 min interval, then mortality rates can be reduced to zero with 100% success in developing SE [5]. The major restriction of this model is its ineffectiveness in mice species. In contrast to rats, lithium pretreatment in mice does not potentiate the convulsive activity of pilocarpine [78].

5.3 Trouble-shootings/ Considerations/ Precautions

1. Animals that do not exhibit SE development feature while seizure scoring should be excluded from further experimental procedures.

2. Anesthetic agents are to be selected wisely to minimize interruptions with the outcomes. Isoflurane is a preferred choice; however, it requires an induction box which may not be available everywhere. Hence, other anesthetic agents such as ketamine and xylazine can be administered as substitutes.

3. Precautions have to be taken while placing the animal on the stereotaxic platform to obtain an accurate position for capillary injections. It is mandatory to check the coordinates properly before injecting. Similarly, the positioning of the electrodes is critical for accurate EEG readings.

4. For EEG recording, the protocol for electrode implantation timings could vary depending on the experimental purposes and the strains of animals employed. For detailed tips on surgical/implantation issues and electrode placement and other artifacts, the readers may refer to Chapter 8.

5. Animals are to be monitored daily and closely to ensure their well-being post-SE. There can be sensitivity reactions to same anesthetic agents for different animals. Also, their health and recovery should be assured before conduct of any other behavioral tests for the neuropsychiatric manifestations associated with epilepsy.

6. A common but crucial problem occurs due to clogging of the small diametric glass pipette and/or due to the capillary suction of blood or other fluids into the capillary injection tube. Hence, caution is advised to check the injecting capillary and glass pipette before administering.

7. Direct administration of pilocarpine without pretreatment with anticholinergic drugs might worsen the condition of experimental animal due to the severe peripheral cholinergic side effects (such as pilocarpine erection, salivation, tremor, chromodacryorrhea, and diarrhea) of pilocarpine [5].

8. Skipping the post-SE treatment with diazepam results in spontaneous remission of seizures 5–6 h after pilocarpine administration, and the animals might die or enter into post-ictal coma, lasting 1–2 days. Therefore, termination of acute seizures by diazepam improves the chances of successful model development.

9. During behavioral seizure scoring, care must be taken to place single animal per cage to avoid animals hurting each other in severe seizure conditions.

10. SE induction by pilocarpine results in the reduction of animal body weight (10–20%). Hence, daily measurement of body weight is necessary until animals start gaining weight and eating moist diet.

11. Survival rate could be improved by helping animal in recovering after pilocarpine treatment by giving 5% dextrose which provides energy and hydration to animals unable to consume food for few days after SE.

12. Lithium-pilocarpine combination should not be given in mice because they are resistant to the potentiating effect of lithium.

6   Conclusion

Even after decades of active epilepsy research and despite the availability of various other models, the characteristic features of which are depicted in Table 2, the kainic acid model and pilocarpine model are still considered one of the most widely used methods for the post-status epilepticus induction of TLE. In order to understand the complicated aspects of human TLE, researchers utilize these models to investigate chronic epileptic seizures. These animal

Table 2
Characteristics of other models available for inducing TLE for rodents

Animal models	Proposed mechanism of action	Induction method and dose	Seizure type	Mortality	SRS	Limitations
OP pesticide model	Inhibits acetylcholinesterase Increases glutamate release	Intrahippocampal/peripheral injection: diisopropylfluorophosphate (DFP; 1.25–4 mg/kg, s.c.), paraoxon (POX; 0.1–4 mg/kg, s.c.), and soman (110 μg/kg, s.c.)	Limbic SE	0–12%	Present	Toxic chemicals associated with ↑ mortality
Flurothyl model	Sodium (Na⁺) channel opening, inhibition of GABA synthesis	Inhaled by rats at 40 μL/min	Tonic-clonic seizures	6%	Absent in adults	Evaporation rate fluctuates with varied atmospheric conditions
Electrical model (PPS)	Activation of hippocampal afferent pathways	10–12-day rat pups stimulated at 20 Hz for 0.2–0.4 ms	Complex partial seizures develop TLE	18–40%; <5% for older animals	Present	Labor intensive; complex procedure
Hypoxia/ischemia model	Reduced blood O$_2$ level resulting in acidosis-induced seizures	Rat pups exposed to 4–7% of oxygen for 15 min	Repeated short tonic-clonic seizures develop perinatal brain injury	20%	Present	Seizure susceptibility varies with the strain and age of rodents; neuron loss develops beyond temporal lobe
Hyperthermia model	Increased body temperature induces febrile seizures	Immature rodents' body temperature raised up to 41 °C through stream of heated air	Febrile seizures characterized by myoclonic jerks	10%	Present	Possibility of mortality due to heat exposure

(continued)

Table 2
(continued)

Animal models	Proposed mechanism of action	Induction method and dose	Seizure type	Mortality	SRS	Limitations
Fluid percussion injury model	Development of post-traumatic brain injury	Pressurized injection of saline at 2 atm pulse against cranial dura of rats	Generalized tonic-clonic seizures develop TLE	10%	Present	Laborious, long latency indistinct seizure origin site
Tetanus toxin	Inhibits the release of GABA and glycine	Intrahippocampal injection at (10–20 mouse lethal doses in <1μL	Complex partial seizure develops TLE	0–10%	Present	Difficult in handling; ↑ mortality
Genetic (audiogenic model)	Seizure-like symptoms induced by high-frequency stimulations	High-intensity acoustic stimulation in genetically prone rats	Wild running and tonic-clonic seizures	0%	Absent	Genetic models are not explored widely in studying TLE

models have allowed investigators to study ictogenesis and epileptogenesis from single neurons to networks, and the results obtained do support the hypothesis that epilepsy results from a complex interaction between aberrant network activity and morphological changes. Regarding the various routes of administration employed by researchers, the intracerebral and systemic procedures yield similar results in terms of latency to SE, behavioral symptoms, duration of the latent period, and electroencephalographic features of the latent and chronic periods. The main difference lies in the extent of neuropathological damage that is induced by each route of administration, with the intracerebral administration producing unilateral temporal lobe lesions as observed in humans with TLE. The choice of either one or the other method will thus depend on the questions that need to be answered: intracerebral administration of KA may be used to investigate the effect of epileptiform activities in circumscribed pathological networks and the effect of seizures on surrounding and presumably healthy tissue, whereas systemic administration should be employed to study the selective vulnerability of multiple brain regions to the agent and to the occurrence of a more widespread epileptic disease. In conclusion, the KA model and pilocarpine model have provided a better understanding of the processes underlying TLE and have important contributions for more efficient and targeted therapeutic drugs not only for seizures but also in search for disease-modifying and anti-epileptogenic therapies.

References

1. Thom M (2014) Hippocampal sclerosis in epilepsy: a neuropathology review. Neuropathol Appl Neurobiol 40(5):520–543

2. Bertoglio D, Amhaoul H, Van Eetveldt A et al (2017) Kainic acid-induced post-status epilepticus models of temporal lobe epilepsy with diverging seizure phenotype and neuropathology. Front Neurol 8:588

3. Trinka E, Cock H, Hesdorffer D et al (2015) A definition and classification of status epilepticus–report of the ILAE Task Force on classification of status epilepticus. Epilepsia 56 (10):1515–1523

4. Lévesque M, Avoli M, Bernard C (2016) Animal models of temporal lobe epilepsy following systemic chemoconvulsant administration. J Neurosci Methods 260:45–52

5. Curia G, Longo D, Biagini G et al (2008) The pilocarpine model of temporal lobe epilepsy. J Neurosci Methods 172(2):143–157

6. Cavalheiro E, Riche D, La Salle GLG (1982) Long-term effects of intrahippocampal kainic acid injection in rats: a method for inducing spontaneous recurrent seizures. Electroencephalogr Clin Neurophysiol 53(6):581–589

7. Lévesque M, Avoli M (2013) The kainic acid model of temporal lobe epilepsy. Neurosci Biobehav Rev 37(10 Pt 2):2887–2899

8. Goffin K, Nissinen J, Van Laere K et al (2007) Cyclicity of spontaneous recurrent seizures in pilocarpine model of temporal lobe epilepsy in rat. Exp Neurol 205(2):501–505

9. Shibley H, Smith BN (2002) Pilocarpine-induced status epilepticus results in mossy fiber sprouting and spontaneous seizures in C57BL/6 and CD-1 mice. Epilepsy Res 49 (2):109–120

10. Jefferys J, Steinhaeuser C, Bedner P (2016) Chemically-induced TLE models: topical application. J Neurosci Methods 260:53–61

11. Chakir A, Fabene PF, Ouazzani R et al (2006) Drug resistance and hippocampal damage after delayed treatment of pilocarpine-induced epilepsy in the rat. Brain Res Bull 71 (1–3):127–138

12. Smyth MD, Barbaro NM, Baraban SC (2002) Effects of antiepileptic drugs on induced epileptiform activity in a rat model of dysplasia. Epilepsy Res 50(3):251–264

13. Glien M, Brandt C, Potschka H et al (2002) Effects of the novel antiepileptic drug levetiracetam on spontaneous recurrent seizures in the rat pilocarpine model of temporal lobe epilepsy. Epilepsia 43(4):350–357

14. Grabenstatter HL, Clark S, Dudek FE (2007) Anticonvulsant effects of carbamazepine on spontaneous seizures in rats with Kainate-induced epilepsy: comparison of intraperitoneal injections with drug-in-food protocols. Epilepsia 48(12):2287–2295

15. Löscher W (2017) Animal models of drug-refractory epilepsy. In: Models of seizures and epilepsy. Elsevier, London, pp 743–760

16. Shinozaki H, Konishi SJBR (1970) Actions of several anthelmintics and insecticides on rat cortical neurones. Brain Res 24(2):368–371

17. Vincent P, Mulle CJN (2009) Kainate receptors in epilepsy and excitotoxicity. Neuroscience 158(1):309–323

18. Thom M (2014) Review: hippocampal sclerosis in epilepsy: a neuropathology review. Neuropathol Appl Neurobiol 40(5):520–543

19. Dedeurwaerdere S, Fang K, Chow M et al (2013) Manganese-enhanced MRI reflects seizure outcome in a model for mesial temporal lobe epilepsy. NeuroImage 68:30–38

20. Fisahn A, Contractor A, Traub RD et al (2004) Distinct roles for the kainate receptor subunits GluR5 and GluR6 in kainate-induced hippocampal gamma oscillations. J Neurosci Off J Soc Neurosci 24(43):9658–9668

21. Mulle C, Sailer A, Perez-Otano I et al (1998) Altered synaptic physiology and reduced susceptibility to kainate-induced seizures in GluR6-deficient mice. Nature 392 (6676):601–605

22. Sabilallah M, Fontanaud P, Linck N et al (2016) Evidence for status epilepticus and pro-inflammatory changes after intranasal Kainic acid administration in mice. PLoS One 11 (3):e0150793

23. Bielefeld P, Sierra A, Encinas JM et al (2017) A standardized protocol for stereotaxic intrahippocampal administration of Kainic acid combined with electroencephalographic seizure monitoring in mice. Front Neurosci 11:160

24. Ben-Ari Y (1985) Limbic seizure and brain damage produced by kainic acid: mechanisms and relevance to human temporal lobe epilepsy. Neuroscience 14(2):375–403

25. Sharma AK, Reams RY, Jordan WH et al (2007) Mesial temporal lobe epilepsy: pathogenesis, induced rodent models and lesions. Toxicol Pathol 35(7):984–999

26. Turski WA, Cavalheiro EA, Schwarz M et al (1983) Limbic seizures produced by pilocarpine in rats: behavioural, electroencephalographic and neuropathological study. Behav Brain Res 9(3):315–335

27. Schauwecker PE (2012) Strain differences in seizure-induced cell death following pilocarpine-induced status epilepticus. Neurobiol Dis 45(1):297–304

28. Blumcke I, Thom M, Aronica E et al (2013) International consensus classification of hippocampal sclerosis in temporal lobe epilepsy: a Task Force report from the ILAE Commission on Diagnostic Methods. Epilepsia 54 (7):1315–1329

29. Hamilton SE, Loose MD, Qi M et al (1997) Disruption of the m1 receptor gene ablates muscarinic receptor-dependent M current regulation and seizure activity in mice. Proc Natl Acad Sci U S A 94(24):13311–13316

30. Marchi N, Oby E, Batra A et al (2007) In vivo and in vitro effects of pilocarpine: relevance to ictogenesis. Epilepsia 48(10):1934–1946

31. Furtado Mde A, Braga GK, Oliveira JA et al (2002) Behavioral, morphologic, and electroencephalographic evaluation of seizures induced by intrahippocampal microinjection of pilocarpine. Epilepsia 43(Suppl 5):37–39

32. Castro OW, Furtado MA, Tilelli CQ et al (2011) Comparative neuroanatomical and temporal characterization of FluoroJade-positive neurodegeneration after status epilepticus induced by systemic and intrahippocampal pilocarpine in Wistar rats. Brain Res 1374:43–55

33. Furtado MA, Castro OW, Del Vecchio F et al (2011) Study of spontaneous recurrent seizures and morphological alterations after status epilepticus induced by intrahippocampal injection of pilocarpine. Epilepsy Behav: E&B 20 (2):257–266

34. Turski WA, Cavalheiro EA, Turski L et al (1983) Intrahippocampal bethanechol in rats: behavioural, electroencephalographic and neuropathological correlates. Behav Brain Res 7 (3):361–370

35. Buckmaster PS, Haney MM (2012) Factors affecting outcomes of pilocarpine treatment in a mouse model of temporal lobe epilepsy. Epilepsy Res 102(3):153–159

36. Racine RJ (1972) Modification of seizure activity by electrical stimulation. II. Motor seizure. Electroencephalogr Clin Neurophysiol 32 (3):281–294

37. Jefferys J, Steinhauser C, Bedner P (2016) Chemically-induced TLE models: topical application. J Neurosci Methods 260:53–61

38. McLin JP, Steward O (2006) Comparison of seizure phenotype and neurodegeneration induced by systemic kainic acid in inbred, outbred, and hybrid mouse strains. Eur J Neurosci 24(8):2191–2202

39. McKhann G II, Wenzel H, Robbins C et al (2003) Mouse strain differences in kainic acid sensitivity, seizure behavior, mortality, and hippocampal pathology. Neuroscience 122 (2):551–561

40. Zheng X-Y, Zhang H-L, Luo Q et al (2010) Kainic acid-induced neurodegenerative model: potentials and limitations. Biomed Res Int 2011:457079

41. Arcieri S, Velotti R, Noè F et al (2014) Variable electrobehavioral patterns during focal nonconvulsive status epilepticus induced by unilateral intrahippocampal injection of kainic acid. Epilepsia 55(12):1978–1985

42. Puttachary S, Sharma S, Tse K et al (2015) Immediate epileptogenesis after kainate-induced status epilepticus in C57BL/6J mice: evidence from long term continuous video-EEG telemetry. PLoS One 10(7):e0131705

43. Van Nieuwenhuyse B, Raedt R, Sprengers M et al (2015) The systemic kainic acid rat model of temporal lobe epilepsy: long-term EEG monitoring. Brain Res 1627:1–11

44. Carriero G, Arcieri S, Cattalini A et al (2012) A Guinea pig model of mesial temporal lobe epilepsy following nonconvulsive status epilepticus induced by unilateral intrahippocampal injection of kainic acid. Epilepsia 53 (11):1917–1927

45. Valente SG, Naffah-Mazzacoratti MG, Pereira M et al (2002) Castration in female rats modifies the development of the pilocarpine model of epilepsy. Epilepsy Res 49(3):181–188

46. Tse K, Puttachary S, Beamer E et al (2014) Advantages of repeated low dose against single high dose of kainate in C57BL/6J mouse model of status epilepticus: behavioral and electroencephalographic studies. PLoS One 9 (5):e96622

47. Yang Q, Huang Z, Luo Y et al (2019) Inhibition of Nwd1 activity attenuates neuronal hyperexcitability and GluN2B phosphorylation in the hippocampus. EBioMedicine 47:470–483

48. Kienzler-Norwood F, Costard L, Sadangi C et al (2017) A novel animal model of acquired human temporal lobe epilepsy based on the simultaneous administration of kainic acid and lorazepam. Epilepsia 58(2):222–230

49. Paxinos G, Watson C (2006) The rat brain in stereotaxic coordinates: hard cover edition. Elsevier, Amsterdam

50. Gurbanova AA, Aker RG, Sirvanci S et al (2008) Intra-amygdaloid injection of kainic acid in rats with genetic absence epilepsy: the relationship of typical absence epilepsy and temporal lobe epilepsy. J Neurosci 28 (31):7828–7836

51. Raedt R, Van Dycke A, Van Melkebeke D et al (2009) Seizures in the intrahippocampal kainic acid epilepsy model: characterization using long-term video-EEG monitoring in the rat. Acta Neurol Scand 119(5):293–303

52. Mouri G, Jimenez-Mateos E, Engel T et al (2008) Unilateral hippocampal CA3-predominant damage and short latency epileptogenesis after intra-amygdala microinjection of kainic acid in mice. Brain Res 1213:140–151

53. Zeidler Z, Brandt-Fontaine M, Leintz C et al (2018) Targeting the mouse ventral hippocampus in the intrahippocampal kainic acid model of temporal lobe epilepsy. eNeuro 5(4):e0158 1–16

54. Chen Z, Ljunggren H-G, Bogdanovic N et al (2002) Excitotoxic neurodegeneration induced by intranasal administration of kainic acid in C57BL/6 mice. Brain Res 931 (2):135–145

55. Lima IVA, Campos AC, Bellozi PMQ et al (2016) Postictal alterations induced by intrahippocampal injection of pilocarpine in C57BL/6 mice. Epilepsy Behav: E&B 64 (Pt A):83–89

56. Cifelli P, Grace AA (2012) Pilocarpine-induced temporal lobe epilepsy in the rat is associated with increased dopamine neuron activity. Int J Neuropsychopharmacol 15(7):957–964

57. Medina-Ceja L, Pardo-Pena K, Ventura-Mejia C (2014) Evaluation of behavioral parameters and mortality in a model of temporal lobe epilepsy induced by intracerebroventricular pilocarpine administration. Neuroreport 25 (11):875–879

58. Borges K, Gearing M, McDermott DL et al (2003) Neuronal and glial pathological changes during epileptogenesis in the mouse pilocarpine model. Exp Neurol 182(1):21–34

59. Paxinos G, Franklin KB (2019) Paxinos and Franklin's the mouse brain in stereotaxic coordinates. Academic Press, Massachusetts

60. De Biase D, Paciello O (2015) Essential and current methods for a practical approach to comparative neuropathology. Folia Morphol (Warsz) 74(2):137–149

61. Paxinos G, Watson C (2004) The rat brain in stereotaxic coordinates–the new coronal set, 5th edn

62. Ben-Ari Y, Tremblay E, Riche D et al (1981) Electrographic, clinical and pathological alterations following systemic administration of kainic acid, bicuculline or pentetrazole: metabolic mapping using the deoxyglucose method with special reference to the pathology of epilepsy. Neuroscience 6(7):1361–1391

63. Medvedev A, Mackenzie L, Hiscock J et al (2000) Kainic acid induces distinct types of epileptiform discharge with differential involvement of hippocampus and neocortex. Brain Res Bull 52(2):89–98

64. Lévesque M, Langlois JP, Lema P et al (2009) Synchronized gamma oscillations (30–50 Hz) in the amygdalo-hippocampal network in relation with seizure propagation and severity. Neurobiol Dis 35(2):209–218

65. Bragin A, Azizyan A, Almajano J et al (2009) The cause of the imbalance in the neuronal network leading to seizure activity can be predicted by the electrographic pattern of the seizure onset. J Neurosci 29(11):3660–3671

66. Hasegawa D, Orima H, Fujita M et al (2002) Complex partial status epilepticus induced by a microinjection of kainic acid into unilateral amygdala in dogs and its brain damage. Brain Res 955(1–2):174–182

67. Nadler JV, Cuthbertson GJ (1980) Kainic acid neurotoxicity toward hippocampal formation: dependence on specific excitatory pathways. Brain Res 195(1):47–56

68. Magloczky Z, Freund T (1993) Selective neuronal death in the contralateral hippocampus following unilateral kainate injections into the CA3 subfield. Neuroscience 56(2):317–335

69. Jefferys J, Steinhäuser C, Bedner P (2016) Chemically-induced TLE models: topical application. J Neurosci Methods 260:53–61

70. Shinoda S, Araki T, Lan JQ et al (2004) Development of a model of seizure-induced hippocampal injury with features of programmed cell death in the BALB/c mouse. J Neurosci Res 76 (1):121–128

71. Ben-Ari Y, Tremblay E, Ottersen O et al (1980) The role of epileptic activity in hippocampal and 'remote' cerebral lesions induced by kainic acid. Brain Res 191(1):79–97

72. Drexel M, Preidt AP, Sperk G (2012) Sequel of spontaneous seizures after kainic acid-induced status epilepticus and associated neuropathological changes in the subiculum and entorhinal cortex. Neuropharmacology 63 (5):806–817

73. White HS, Löscher W (2014) Searching for the ideal antiepileptogenic agent in experimental models: single treatment versus combinatorial treatment strategies. Neurotherapeutics 11 (2):373–384

74. Imran I, Hillert MH, Klein J (2015) Early metabolic responses to lithium/pilocarpine-induced status epilepticus in rat brain. J Neurochem 135(5):1007–1018

75. Marchi N, Fan Q, Ghosh C et al (2009) Antagonism of peripheral inflammation reduces the severity of status epilepticus. Neurobiol Dis 33 (2):171–181

76. Clifford DB, Olney JW, Maniotis A et al (1987) The functional anatomy and pathology of lithium-pilocarpine and high-dose pilocarpine seizures. Neuroscience 23(3):953–968

77. Scholl EA, Dudek FE, Ekstrand JJ (2013) Neuronal degeneration is observed in multiple regions outside the hippocampus after lithium pilocarpine-induced status epilepticus in the immature rat. Neuroscience 252:45–59

78. Groticke I, Hoffmann K, Loscher W (2007) Behavioral alterations in the pilocarpine model of temporal lobe epilepsy in mice. Exp Neurol 207(2):329–349

Chapter 8

Seizure Monitoring in Rodents

Edward H. Bertram

Abstract

There are a growing number of rodent models of epilepsy that are now available. These models have a number of potential research uses, including understanding the process of epileptogenesis, determining the natural history of the disease, and discovering new therapies. Central to these questions is seeing the seizures behaviorally and physiologically. Because the seizures occur spontaneously and unpredictably, continuous EEG and video monitoring over days, weeks, and sometimes months are needed. Recording from rodents over this period of time requires a specialized system that connects the animals to the EEG machine while allowing the animals humane freedom in a designated space. In this chapter we describe the different uses of EEG monitoring as well as provide an outline of the key components of the recording system, including housing, that are needed to perform the studies. In addition, we offer suggestions what precautions are necessary and the potential technical and interpretive pitfalls.

Key words Rodent EEG, Animal epilepsy models, Electrode implantation, In vivo electrophysiology, Epilepsy physiology

1 Introduction

This chapter is about performing EEGs in rodents to document seizures in models of epilepsy. It is a very labor-intensive effort, but these studies are required to document spontaneous seizures which occur randomly and irregularly. EEG may also help in the analysis of induced seizures such as in kindling. In many spontaneous seizure models, there can often be many days between each seizure, so continuous recordings over days and weeks are often necessary to document the seizures and to determine if a treatment has an effect.

In the almost 30 years since the potential for prolonged EEG recording in rats was first described [1], many laboratories and vendors have provided a number of technical options that best fit an investigator's needs. In this chapter, we will describe some approaches with a view toward those needs. We will also touch on issues that the investigator should consider before deciding on a particular monitoring approach. This chapter is about the principles

Divya Vohora (ed.), *Experimental and Translational Methods to Screen Drugs Effective Against Seizures and Epilepsy*, Neuromethods, vol. 167, https://doi.org/10.1007/978-1-0716-1254-5_8,
© Springer Science+Business Media, LLC, part of Springer Nature 2021

Table 1
Reasons for seizure monitoring in rodents

Documentation of epilepsy/seizures
Frequency of seizures
Behavioral nature of seizures
Differentiating seizure-related from non-seizure-related behaviors
Severity of seizures
Duration of seizures
Effect of therapy
Electrophysiology of the seizure
Functional anatomy of the seizure

that should be considered in choosing the approach (technical and experimental protocol) most appropriate for the project. The presented information is not a complete listing. It is a presentation of potential solutions to common problems. We will also describe some potential pitfalls that must be considered.

Some of the more common reasons for monitoring rodent EEG for seizures are outlined in Table 1. Defining the goals for a particular project is important because it will guide the methodology and technology choice as well as the overall effort by laboratory personnel, which is often underestimated. The process of reviewing the data, recording the findings, and maintaining the recording system can easily take half a day alone, and if the recordings are performed 7 days a week, there may need to be an investment in additional personnel.

We will not address the issue of what is a seizure. At present, there is great controversy with regard to EEG patterns that have been reported in some models. The basic issue is that some patterns called seizures are rhythmic activity with spike-like components that are seen in controls as well as in animals that have undergone a procedure intended to induce epilepsy. At present, until the issue is resolved, it is recommended that the EEGs are reviewed blinded first, to identify the pattern in question and then determine if there is a difference between the groups with regard to the appearance of the pattern. In genetic models, the issue of controls for comparison is problematic, as normal cortical activity can vary from strain to strain. If a normal, rhythmic pattern is misinterpreted as a seizure, one can artificially create an epilepsy model when one is really seeing a normal variation in EEG that is under genetic control.

2   Developing the Right Monitoring Protocol

Keep the goals and needs of your project in mind so that the equipment and monitoring protocols are appropriate. Monitoring is very labor intensive, so choose the approach that answers the question efficiently.

Issues that EEG monitoring can help address:

2.1   Does This Animal Have Epilepsy?

If an animal has behaviors that could be seizures, monitor the animal at any time when it is likely to have the spells in question. To assure that abnormal activity is not artifact, video behavioral correlation is essential. It has been our experience that some repetitive, stereotyped, and, at times, bizarre behaviors have no EEG correlation that would indicate seizure. On the other hand, there are times when what appear to be clear ictal discharges on EEG turn out to be artifact from the environment or from some normal activity such as head scratching. The monitoring can be relatively brief to document the presence of seizures if they are frequent but can be lengthy to document infrequent seizures.

2.2   The Natural History of Epilepsy in a Model

There are many questions tied to this goal. When do the seizures start? Is it at a specific age for a genetic model? Do the seizures begin after an injury or other intervention? Are the seizures limited to specific ages? Choose the time before the seizures might begin. Although monitoring without interruption would be ideal, it can be so effort intense that monitoring for blocks of time, with interruptions for several weeks or a month, is adequate to define a time window when the seizures begin and if they are permanent. Such a protocol will also identify the temporal distribution of seizures (daily, multiple daily, clusters of seizures with long periods without). Knowing the natural history will also help determine how suitable a model is for drug screening (animals with few spontaneous seizures generally are not suitable).

2.3   Determining the Effect of Therapy

The design of the testing protocol will depend on the seizure pattern and the availability of the test therapy as well as the half-life of the therapy. With limited amounts of a drug, the seizures must be consistently frequent so that one or two doses will be able to show an effect. As an effective dose is unknown, a dose ranging protocol is necessary, especially if the first dose is ineffective. Pick a set time of the day to give the drug when the animal has seizures over a few hours and give the drugs at the same time. Alternating drug days with placebo days will allow the animal to return to baseline. For animals with less frequent and more unpredictable seizures, longer duration protocols are necessary, and multiple administrations of drug are needed to maintain therapeutic levels over multiple days. Because the therapeutic half-lives of the test drugs can be short, multiple doses in a day may be required, and, because the seizures are less regular with some naturally seizure-free days, drug administration will require multiple animals over multiple days to determine reliably whether the drug has an effect. These protocols will often require amounts of an experimental drug that will be difficult to obtain. Figure 1 presents the different protocols graphically.

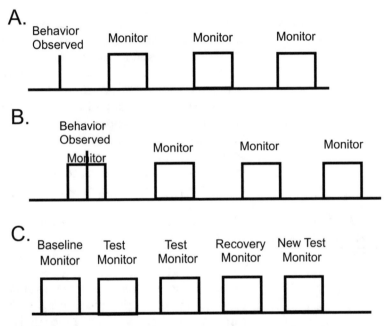

Fig. 1 Monitoring protocols. (**a**) Determining whether the animal has seizures. Monitoring can begin any time after a behavior suggesting seizures is observed. Monitoring blocks may be a few days or several weeks. If none are seen in the first monitoring period, the animal can be monitored at intervals. Because seizures are never recorded, it does not exclude the possibility of infrequent seizures. Epilepsy can only be proved, not disproved. (**b**). Natural history of seizures. Monitoring should begin before the first seizures are suspected to define the time window for the onset of seizures. To determine whether seizures increase in number and behavioral severity over time and when, if ever, the seizures remit, periods of monitoring interspersed with rest periods is efficient. It is possible to limit any one monitoring group to a single monitoring period and categorize the seizure features by age rather than sequential evolution. (**c**). Therapy testing protocol for animals with established seizures. The monitoring periods can be adjusted to the frequency of the seizures and the type of treatment. They can be as short as 1 day or a week or longer depending on the type of treatment. In some cases, there will be a recovery period between each test period, and in some cases, there will be none as the doses are increased sequentially. If animals are to be exposed to multiple drugs, a recovery phase at the end of treatment to determine if the baseline frequency is stable is required

3 Materials

Equipment:

The technical issues of obtaining a quality EEG from multiple animals for days or weeks are challenging. Connecting multiple individuals to a single recording device can reduce the quality of the signal to the point that the data are unusable. Further, maintaining animals on a recording system for prolonged periods of time requires a design that allows for relatively normal activity levels while maintaining a good connection to the EEG. The recording setups, the connections, and housing must be designed with these issues in mind.

Key Components of a Rodent EEG Monitoring System:

1. EEG recording system.
    (a) Hard wired.
    (b) Wireless telemetry.
2. Video recording system.
    (a) Cameras.
    (b) Lighting for nighttime recording.
3. Connections between the animal and the EEG system (hardwired).
    (a) Electrodes.
    (b) Connector between electrodes and cables.
    (c) Electrical swivels to allow full movement of the animal.
4. Housing to allow full animal movement with good visualization of behavior.

4 Methods, Key Considerations, and Troubleshooting

4.1 EEG Recording Systems

4.1.1 Hardwired EEG

For hardwired EEG, a cable connects the animal to the EEG machine. Almost all EEG systems today are digital and come as a complete package with amplifiers, analog to digital converters, digital filters, software and, often, synchronized digital video. The advantage to the system is that it is all integrated which makes set up easier. These digital systems typically record from all electrodes (in the case of multiple animals, all animals) against a single common reference. This requirement raises concerns about the ability to record multiple independent animals against a single reference electrode and still maintain a reasonable recording quality. Because system input resistance often approaches a gig ohm, slight mismatches in electrode impedance across the animals do not cause a major degradation in recording quality which happens with systems with lower input resistances. Systems designed for clinical use as well as systems designed specifically for recording from rodents will work well.

The pairing of electrodes for display on the screen in a digital system is a mathematical calculation that is performed by the machine's computer. The digital approach has the advantage in that many possible electrode pairings can be created from the same primary data.

Helpful Observation:

Although this technique works quite well when only one animal is being recorded, the common reference electrode for many rats is a significant problem when trying to record and create montages from multiple animals at one time. The use of a common reference from a single animal when recording from many can introduce significant amounts of noise into the system, but linking the reference electrode from each of the animals together creates a common reference that is shared by all. This tactic and the grounding of cages or cage bottoms (see below under cages for more details) can allow good recordings from multiple animals on a digital machine.

4.1.2 Radiotelemetry EEG

Radiotelemetry uses a wireless connection between the EEG amplifier that is implanted on or in the animal and the recording system.

Wireless transmission of the EEG signal to receivers that are around or below the cage allows the animals to move around the cage unencumbered by a cable. However, the number of channels is limited to usually one or two channels for each animal, although there are transmitters with up to six channels of recording (Epoch from Epitel, Inc.). Transmitters have a finite battery life so that the units have to be replaced at regular intervals, which can get expensive, depending the number of animals one must record.

Vendors for wireless systems include Data Sciences International (DSI), Epitel, Inc. (Epoch), Millar, and TSE (Stellar and Neurologger). Some use a transmitter that is subcutaneous or intraperitoneal, but some are skull mounted. Exposed skull mounting allows monitoring in very young animals in which subcutaneous or intraperitoneal implantation is not possible. Battery life varies from 2 to 6 months, typically, although some for neonates have a life of a few weeks.

Most systems are limited to a single animal in a cage, because the external receiver can only work with a single transmitter, but some of the newer models can record from several animals that are housed together (social housing). With such systems one could also increase the number of channels per animal by using several transmitters in one animal simultaneously.

Fig. 2 Head mounted telemetry transmitters. Head telemetry device. (**a**). Rat pup (P7) with telemetry transmitter. (**b**). Young adult rat with adult transmitter in standard caging (Both Epoch systems are from Epitel)

Most of the vendors have transmitter models that come in different sizes that are designed to use with animals of different sizes down to about 17 g (a mouse) but one (Epoch by Epitel, Inc.) may be used on rat pups as young as P6 (Fig. 2). This system is not disturbed by the mothers when the pups are returned to the litter, so it is possible to record days of EEG while the pup remains with the litter [4]. On occasion, some mothers may damage the transmitter.

Helpful Observation:

1. One limitation of the wireless systems is the potential stereotaxic depth placement of the electrodes. Most of these systems come with the electrode wires attached to the transmitter, with the general practice to insert those wires directly into the desired targets. This approach works well for cortical recordings but is less well suited for accurate placement into deeper structures (e.g., amygdala).

2. Another limitation is the occasional loss of signal that happens if the animal is positioned such that it is blocking the signal from reaching the receiver.

4.2 Digitization Frequency

Many commercial systems come with a standard 200–400 Hz digitization frequency per channel, which is more than adequate for good visualization of basic seizure patterns and activity up to 100 Hz. This range will not be adequate for investigators who wish to study the very high frequency oscillations (200 Hz or higher). In these situations, a digitization rate of a minimum of twice the highest EEG frequency is needed (e.g., if frequencies of 500 Hz are to be recorded, a digitization frequency of at least 1000 Hz to prevent signal aliasing is needed, but even higher digitization rates are better). The only real disadvantage to this higher sampling rate is the increased storage requirements for the collected data, and there may be a greater chance of recording electrical noise. One may still review the data with a narrower bandwidth.

Helpful Observation:

Wireless EEG monitoring generally has a frequency bandwidth of 1–100 Hz, although a few may reach 140 Hz. For most purposes, this frequency range is not a problem for recording seizures and basic EEG patterns unless one wishes to study ultra-high frequency (anything above 100 Hz for this discussion) in which case a wired system that has amplifiers and digitizers that have an upper limit of 1000 Hz/4000 Hz respectively is essential. There are some transmitters that may record at higher digitization rates, but the battery life falls off rapidly.

4.3   Video Recording

Key considerations for choosing video components:

1. Resolution of video chip.
2. Light sensitivity of camera (night and low light may require infrared sensitivity).
3. Lighting. Placement of light sources and addition of infrared light source.
4. Lens. Manual focus may give more consistent results.

In this section we will describe some of the variations and how the experimental needs can influence equipment choices.

4.3.1   Resolution

Cameras come in many formats and visual resolutions, as well as levels of light sensitivity. In general, it is advisable to use cameras of higher resolution, especially if multiple animals will be recorded simultaneously with a single camera or if there is a need to see subtle behavioral changes. The primary downside to using high definition cameras is the storage requirements for the recordings, which, for color, can easily exceed 2GB/24 h. That requirement can be reduced with the use of black and white, which is usually more than sufficient. Image resolution may also affect the speed of video review. Resolution is not important to determine a behavioral correlate to a particular EEG pattern.

4.3.2   Light Sensitivity

The choice is driven by the expected needs and recording conditions, but if 24-h recordings under light and dark conditions are planned, a camera with infrared sensitivity and infrared illumination is essential to obtain "night time" images.

4.3.3   Lighting

Lighting should be placed so that the cages are well illuminated but with no glare from reflection off the front of the cages. Angling the light from above or the side (or "bouncing" the light off the ceiling) usually works well. A separate source of infrared light on a timer to turn on and off automatically is needed for recordings in the dark period.

4.3.4 Lens

Autofocus is often standard but using a manual zoom lens has some advantages. It is less expensive, and, because the animals are confined to cages, the focus can be set once. Limiting the field of view improves the image quality for review, because one captures just what is needed and nothing else.

4.4 Connecting the Animal to the EEG Machine

In a hardwired system the most critical, yet underappreciated aspect is the distance between the animal's electrodes on its head and the input of the EEG machine. This connection is the most fragile and the source of most of the artifact. The key components are:

1. Connectors: the device that connects the electrodes in the head to the cable.

2. Cable: the bundle of wires that connect the animal to the electrical commutators.

3. Commutators: the electrical swivel that allows the animal full range of movement.

4.4.1 Connectors

These devices hold the electrode tips on the animal side and the receptacles for the electrodes on the cable side. They are matched to allow the two ends to fit together and can be fastened to provide a secure, quiet electrical connection that will not disconnect accidentally. They must be small enough to fit on the animal's head when cemented in place with the acrylate. There should be a position in the connector for each of the electrodes, and the design must allow easy connection of the electrode wire into the positions. The company P1 (formerly Plastics One) has a long history of supplying such connectors that are in a variety of configurations (up to 12 electrode contacts).

Helpful Observation:

Because the placement of the electrodes varies from animal to animal depending on the experiment, there is no standard configuration for the electrodes in the connector. It is essential that the laboratory standardize the position of each electrode in the connector; otherwise it will be impossible to know what any one electrode is recording. Similarly, when creating the connector to the cable side, a consistent placement is required. Careful record keeping of the design of the connector and standardization of the wiring is a must, so that one always knows which connector position relates to which electrode.

4.4.2 Cables to the
Animals

The cables connecting the animal to the commutator are the most vulnerable part of the system. There is a continual wear and tear on the cable, from the activity of the rodent, and animals' tendency to chew makes the connecting cables vulnerable. Because of this fragility, it is important the cable is replaceable (Fig. 3).

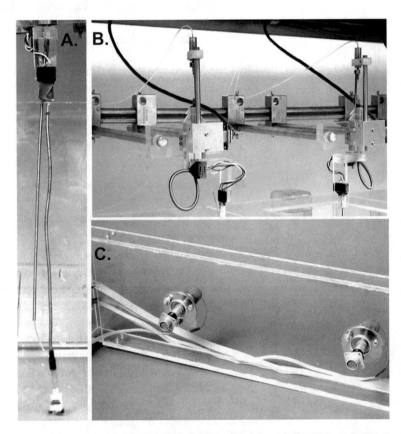

Fig. 3 Cable and commutators. (**a**). Combination of cable for EEG and dialysis tubing. The cable consists of individual wires that are passed through a flexible spring with connectors at both ends. The cable extends to the floor of the cages and is attached to a connector on the commutator assembly. The dialysis tubing is also placed through a spring to prevent chewing by the rat. (**b**). Commutator assembly held by two rods that are held in place by paired rods at the back. Pairing the rods prevents rotation of the assembly. The two rods are attached to an acrylic plate to which the commutator is attached. An acrylic armature at the bottom of the commutator provides a stable connection point for the wires from the commutator to the cable. A fluid swivel is at the top of the assembly and is attached to the commutator by a rod that turns with the assembly. (**c**). Commutators attached to an acrylic shelf that is suspended above the cages. The cables from the commutators to the EEG machine are above the shelf and are protected from the animal

The cable must be flexible so as not to place undue strain on the animal's headset. There are many commercially available cables that come with the connectors attached, but a cheaper alternative uses lighter weight wires (24 gauge or smaller) that are braided and placed through the center of a light-weight door spring. The construction is easy, and the cable is flexible and essentially chew proof. It is important that the individual wires are color coded so that one can trace the connection at the animal end of the cable to the position at the end that connects to the electrical swivels. The length of the cable should be just enough to reach the edges of the cage but not so long that a bend in the cable will allow the animal to chew it.

4.4.3 Electrical Commutators

Electrical commutators or swivels are essential for recordings of more than a few hours. They allow the animal to turn without twisting the cable. In general, there is a center rotating spindle with conductor rings which are in contact with fixed conductor brushes. The primary points that distinguish the commutators of the different vendors are the number of contacts available (4–64), the materials used, the ease of turning, the nature of the cabling connections, and whether the swivel is electrical only, electrical and fluid in one, or electrical but capable of mating with a separate fluid swivel. A key factor is ease of turning: the swivel should take no to minimal effort to turn to reduce headset strain. There are many manufacturers and vendors of commutators including P1, Dragonfly, Crist Instruments, Plexon, Neural Lynx, and Pinnacle Technology. Good quality commutators can be expensive, but they should last for many years. Before making a final purchase, especially if the price is high, see the commutator first.

The commutator can be placed attached to the top of the cage or in a frame above the cage. The cable connection to the animals must be accessible enough to allow easy connections and disconnections. For rats this usually means a connection to the commutator that is at least 30 cm above the bottom of the cage, but generally higher. Part of the positioning is determined by the type of cage and how the commutator and its connections can be accessed. For a cage that can be opened in the front, the swivel could be placed even on top of the cage, whereas for cages open at the top, such as cylinders, the commutator should be placed 5–8 cm above the cage top to give the researcher sufficient working space. Several examples are shown in Fig. 3.

4.5 Cages

The cages should allow unrestricted movement of the animal. There should be no overhead height limitations so that the animal can rear up and explore without hitting its head (and thus electrode assembly) on a cage top. If the video monitoring is to be performed from the side or front, the sides of the cage must be transparent. If video monitoring will be performed from above, the cage tops must

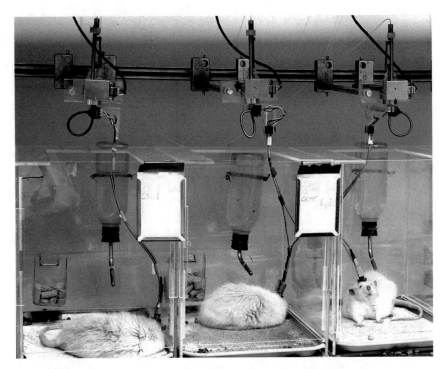

Fig. 4 Multiple rat monitoring cages on metal shelving. Row of cages with overhead electrical and fluid swivels. Cages are built from sheet acrylic and have front opening doors that slide up and down in a groove. In some cases, the front panel can be cut into an upper and lower half so that the top half can be removed to allow access to the animal while still providing a barrier. Depending on regulations, food may be placed in hanging holders (shown) or on the wire mesh flooring. Each cage is 30 cm wide, 45 cm deep, and 45 cm high. In addition, there is a partial opening at the top of the cage that allows access for the cables while providing a barrier around the top edges to reduce the chance for escape

either be clear or absent, and the commutator placed so that it does not block the view (Fig. 4).

Cage dimensions must meet (or preferably exceed) the regulatory requirements for sufficient room to move and turn freely while keeping the space small enough so that the rat never places a significant strain on the cable. Cylindrical cages are ideal but create problems for access to the animals from above and can be challenging if there are fluid lines for infusions or dialysis. They are easy to construct and can be sized appropriately for the animal (Fig. 5).

Fig. 5 Mouse cage. It is made from a 15 cm diameter acrylic tube with a fine mesh wire bottom. The top is made from a funnel with the tip removed. The pins from the mouse are connected to free floating wires that go to the headset. Mice have been monitored continuously for up to several weeks with this or slightly modified cable connections. With the use of i-Bond to create headsets, it may be possible, if the commutator is of sufficiently low turning resistance, to connect the pins of the headset directly to the cable sockets without a formal connector

Helpful Observation:

The electrical environment in and around the cage can influence the recording quality, especially if multiple animals are recorded. Wire mesh bottoms connected to metal shelving ground the cages to one another. Although noise problems from other sources remain a problem, this common grounding of the cages minimizes the electrical isolation of the animals. A wire mesh bottom just above the litter tray, with wires connecting these bottoms to one another and to the underlying metal shelves, significantly reduces the noise. Grounding the cages together is not a problem for wireless systems.

4.6 Setting Up for Recording

This section will describe the process of connecting an animal to the EEG machine:

1. Implanting the electrodes.
2. Securing the electrodes to the skull so that they don't shift.
3. Creating the connection between the animal and the machine.

This section will include general principles as well as a more practical "how to" process. Some of the steps will be modified to fit the specific issues a protocol and the nature of the recording, but the basic principles remain the same. This process requires practice, and consistent results are unlikely until the investigator and technicians have performed the procedures many times. A number of videos demonstrating the basic process are listed in the reference section.

4.6.1 Surgical/ Implantation Issues

There are four basic components for the process: the electrodes for implantation, small stainless steel or plastic screws (0 gauge or smaller) for fixing the headset to the skull, the material for covering the skull and fixing the electrodes in place, and the connector between the electrode and the cable (the headset).

Electrodes:

The electrodes are often fabricated in the laboratory from wire and connector pins, as they are easy to make, inexpensive, and readily customized. Other types of electrodes (e.g., concentric, in which one wire is inside the other) can be purchased and are ready to implant. Lab-built electrodes are single or twisted pair wires (stainless steel or, less commonly, platinum iridium) that are insulated with a thin coating of Teflon. Twisted pair electrodes may provide better recordings of local activity and are generally preferred for focal stimulation, such as in kindling. They are made by taking a strand of the Teflon insulated stainless steel, holding the two ends together with a pair of forceps and twisting the ends while there is a small rod or stick at the loop end. The loop is cut and connecting pins are placed on each end, while the twisted portion is cut to the needed length. The single wires are usually placed in the cortex or subdurally, because they tend to bend and go off course making accurate deeper placement difficult. In placing the electrodes in the superficial cortex when electrical stimulation is considered, avoid exposing bare wire where it might come into contact with the dura, which is quite pain sensitive. "Overtwisting" a twisted pair electrode may result in a break in the insulation, leading to inaccurate recordings or stimulation. Commercial concentric bipolar electrodes are best for recording from well-defined, restricted regions.

Materials for Electrodes and Implantation:

1. 0-gauge jeweler's screws (e.g., plastic or stainless steel).
2. Self-retaining jeweler's screwdriver that holds screws while placing in skull.
3. Electrodes (insulated wire or commercial).
4. Stereotactic frame (e.g., Kopf or Stoelting).

5. High speed drill (e.g., Dremel) and a burr bit (about 1.5 mm in diameter) for electrode insertion and a drill bit (1.0 mm) for drilling holes for the screws).

6. Polymerizing material to hold electrode in place (e.g., dental acrylic).

Implantation Tips and Notes:

There are two issues that are extremely important when performing the surgeries to implant the electrodes for long-term recordings: skull preparation and screw placement.

1. Removing all of the soft tissue from the skull surface improves adhesion of the headset plastic to the skull and minimizes the risk of tissue regrowth under the headset which can, over time, loosen the headset and cause premature loss as well as provide space for an infection to develop. Using gauze moistened with an iodine antiseptic, followed by a similar vigorous scrub with a 2% hydrogen peroxide solution, has been very helpful.

2. The screws should be placed to prevent rocking of the headset either front to back or side to side. Thus, several screws should be placed a few millimeters (or more) from the anterior and posterior borders as well as from the sides. Placing screws on the lateral aspect of the skull can prevent the lateral rocking. The massetter muscle attachments to the lateral ridges of the skull must be lifted up (usually with a very gentle scraping with fine rounded tip forceps to form a small pocket between the muscle and the skull. The screw is placed a millimeter or two below the lateral ridge, and the liquid plastic dripped into the space, covering the screw.

3. Once the screws are in place, one can place the electrodes. For cortical only recordings, drill the holes with the burr at the desired locations, and if recording monopolar electrodes, drill another over an electrically neutral site (over the frontal sinus or the cerebellum) for a reference electrode. If deeper intracerebral recordings are needed, develop the stereotactic coordinates with a stereotactic atlas [3]. Placement of these electrodes requires a stereotactic frame. If multiple electrodes are being placed, each one must be secured in place with the acrylic that is allowed to harden before the next electrode is inserted.

To see a video on implantation surgery in rats, visit this Jove article that is open access (https://www.jove.com/video/3565/surgical-implantation-chronic-neural-electrodes-for recording-single) [2]. Other videos are available to help a laboratory develop its own methods.

1. Prepare skull surface with antiseptic solution followed by a gentle scrub with 2% hydrogen peroxide.

2. Drill holes for skull screws with sterile 1 mm drill bit.

3. Place screws in each of the holes. Insert so that 0.5–1 mm of screw is below the skull inner table. Gently lift masseter muscle away from skull for lateral screws.

4. Use 1.5 mm burr bit to drill holes for the electrode(s). Use stereotactic arm to identify anterior-posterior and medial lateral position of hole.

5. Insert electrode to appropriate depth. After each insertion place enough acrylate to hold the electrode in place and allow to harden before moving to next electrode.

6. Before placing connector, make sure that the acrylate has hardened and covers the screws and electrode insertion points in a continuous manner and that electrodes are fixed in place.

Connectors:

Connectors organize and hold the ends of the electrodes coming out of the animal and connect those electrodes to a cable that goes to the recording system. There are a variety of commercially available connectors and the ones from P1 Technologies (formerly Plastics One) are used by a number of laboratories. They are complete systems and have a secure connection between the headset and the cable. In connecting the animals, it is usually less stressful for the animal, investigator, and connector to anesthetize the animals lightly with an inhalational anesthetic before attempting to attach the cable.

Headset Resin:

Dental methacrylate is the traditional material for headsets because it can harden in 5–10 min to allow rapid completion of the headset. A variation of the resin is i-Bond from Heraeus-Kulzer. With a polymerization lamp, the resin cures rapidly and can be applied in layers. i-Bond bonds well to the skull surface, but good skull surface preparation is still recommended.

4.7 Age- and Species-Specific Issues

The basic issues for monitoring (EEG recordings, video monitoring) are the same regardless of the species or age of the animal. There are differences, however, in how one attaches a 25 g mouse to the system as compared to a 500 g rat. The rat skull is thick enough so that the headset can withstand movement a cable attached to a swivel. For the mouse, there can be no significant strain on the headset. In addition, the mouse has much less room on its skull for screws and electrodes. For mice, the commutator must have little or no resistance to turning, and the cable should place minimal to no weight on the animal. Instead of a connector, the electrode wires can be connected directly into the sockets on the cable (using forceps) and the friction holds them in place. The loose electrode wires reduce strain on the headset but are stiff enough to turn a low resistance swivel. Quality recordings over a number of weeks have been possible with this approach.

4.8   Electrical Artifact   The greatest problem in recording multiple animals (or even a single animal) is electrical artifact which can come from a variety of sources. Although it is usually quite obvious, completely hiding the underlying EEG signal, it can also mimic interictal or ictal activity to the point that it is impossible to tell real brain activity from signal that originates outside the skull (Fig. 6). One must always be aware of this potential and be suspicious of what the

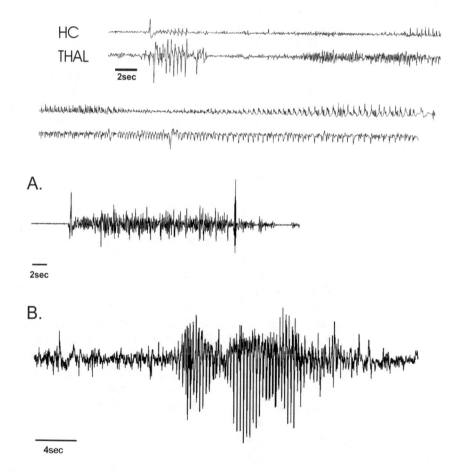

Fig. 6 Seizure vs. artifact. The top EEG is a seizure recorded from the hippocampus and midline thalamus. There is an initial spike or spikes, followed by a period of relative suppression and the build-up of repetitive spiking that slows before stopping. (**a**). Artifact from a rat with a broken cable wire from movement. There is an initial spike, followed by repetitive spiking that also slows before stopping. (**b**). Artifact from head scratching. There is repetitive spiking that waxes and wanes. The only way to distinguish among the three is through the video data

recording really shows. In this section, we will outline the common and some of the less common causes of artifact and how to diagnose and fix the problem.

Perhaps the most common source of artifact, especially if it involves a single animal, is a loose connection or broken wire. A key tool for locating the source of this type of artifact is an ohmmeter for measuring resistance at selected points in the lines that connect the animals to the machine. A good ohmmeter with a 9 V battery such as a Simpson Model 260 is necessary when measuring intracerebral electrode resistance which normally ranges between 5000 and 20,000 Ohms with normal stainless-steel electrodes. The higher voltage of this ohmmeter is necessary to obtain reliable and consistent readings in a volume conductor (the brain). Measuring resistance in a logical sequence from the level of the electrode pins, the commutator channels and at the cables for the continuity of individual wires, and potential shorting between wires should be a routine procedure to identify the source or artifact.

As mentioned earlier, electrically isolating animals in cages can contribute to noise. Using cages with wire mesh bottoms and grounding the metal components of the cages to the shelving can help. If there is metal housing to the electrical swivels, it may be helpful to link those casings to a common ground. Finally, link all implanted animal ground electrodes to a single common ground input in the EEG machine. This approach will usually reduce the noise. Incomplete grounding of cages and animals is often obvious because the problem is seen in all channels, but sometimes the source of the noise is not associated with the channel(s) that is causing it, as the noise may appear in all of the channels except the one(s) associated with the real cause.

When confronted with noise, think first about the common sources such as bad grounding, loose connections (between connectors, between pins and wires anywhere from the intracerebral electrodes to the pins going to the EEG input and poor connections inside the commutator), animal activity (such as head scratching or chewing), ungrounded cages, and unmatched impedances among the intracerebral electrodes. Some less commonly recognized sources of noise include open (and turned on) EEG channels with crosstalk to other channels, bad connectors in the headbox, and crosstalk at the digitizer. Localizing the problem can take time and requires a logical stepwise approach. In general, the first thing to consider when seeing intriguing and potentially exciting recordings is that one is seeing noise. Only after one has excluded noise as a source can one feel a little excited about the results on the EEG. Remembering the admonition that all EEG is artifact until proved otherwise will frequently prevent gross misinterpretations of data.

References

1. Bertram EH, Williamson JM, Cornett JC, Spradlin S, Chen ZF (1997) Design and construction of a long-term continuous video-EEG monitoring unit for simultaneous recording of multiple small animals. Brain Res Protocol 2:85–97

2. Gage GJ, Stoetzner CR, Richner T, Brodnick SK, Williams JC, Kipke DR (2012) Surgical implantation of chronic neural electrodes for recording single unit activity and electrocorticographic signals. JOVE J. https://doi.org/10. 3791/3565. https://www.jove.com/t/3565/surgical-implantation-chronic-neural-electrodes-for-recording-single

3. Paxinos G, Watson C (2013) The Rat Brain in Stereotaxic coordinates, 7th edn. Academic, London

4. Zayachkivsky A, Lehmkuhle MJ, Fisher JH, Ekstrand JJ, Dudek FE (2013) Recording EEG in immature rats with a novel miniature telemetry system. J Neurophysiol 109:900–911

Part IV

Models for Refractory Seizures

Chapter 9

Protocol for 6 Hz Corneal Stimulation in Rodents for Refractory Seizures

Razia Khanam and Divya Vohora

Abstract

The 6-hertz (6 Hz) model is an effective tool to screen drugs for psychomotor seizures and/or treatment-resistant focal seizures. It is known to identify the drugs or molecules acting via different mechanisms, unlike the traditional models. Seizure is induced by corneal stimulation (6 Hz, 0.2 ms for 3 s duration) using electroconvulsometer. The characteristic features seen in rodents are stun position and minimal clonic phase, followed by stereotyped behavior observed for 120 s, at stimulus of different intensities like 22, 32, or 44 mA. The animals are considered as protected if no stereotype behavior is observed and the animal resumes normal exploratory behavior within 10 s. Despite being effective in identifying the antiseizure potential of certain drugs possessing newer mechanisms, the model cannot correlate the neuronal and pathophysiological adaptations occurring in chronic cases of epilepsy, which is considered as its drawback.

Key words 6 Hz, 6-hertz psychomotor seizures, Resistant seizures, Levetiracetam, Valproate

---

1 The 6-Hertz Model of Psychomotor Seizures

The mouse 6-hertz (6 Hz) model, originally developed in the 1950s, was later rediscovered as an effective tool to screen drugs for psychomotor seizures and/or treatment-resistant focal seizures. It was also termed as minimal electroshock seizure threshold test [1] and was utilized to screen established as well as novel compounds that may have antiseizure potential [2, 3]. The compounds showing antiseizure activity via different mechanisms (like sodium channel blockade, GABAergic potentiation, modulation of synaptic vesicle glycoprotein 2A 2 (SV2A), modification of potassium channels, etc.) show different grades of efficacy in this model [4]. Initially, the use of this model in the primary screening of antiseizure drugs (ASDs) was not recommended due to its lack of ability to screen phenytoin, however, several years later.

The test is a simple screening tool considered useful in determining the efficacy of a new antiseizure drug for treatment-resistant

Divya Vohora (ed.), *Experimental and Translational Methods to Screen Drugs Effective Against Seizures and Epilepsy*, Neuromethods, vol. 167, https://doi.org/10.1007/978-1-0716-1254-5_9,
© Springer Science+Business Media, LLC, part of Springer Nature 2021

focal seizures, a characteristic that makes it distinct from traditional maximal electroshock seizure (MES) and s.c. pentylenetetrazole (PTZ) models. Though the traditional screening tests like MES and s.c. PTZ are also extremely useful in determining antiseizure effectiveness against generalized tonic-clonic seizures (GTCS) or generalized myoclonic seizures, they may miss the identification of effectiveness against refractory focal seizures [5]. Thus, an investigational drug or a lead compound is required to be screened in this model also for its ability to show efficacy in the treatment-resistant seizures (if any), even if they fail in the MES or PTZ models [3]. For instance, levetiracetam, an established broad-spectrum antiseizure drug, was found ineffective in standard MES and PTZ tests in rats and mice initially, up to the doses of 500 mg/kg intraperitoneally [6, 7]. Later, however, it was found to be effective in blocking 6 Hz seizures in epileptic mice [3, 8] and in suppressing seizures in audiogenic DBA/2 mice [9]. Thus, the antiseizure potential of levetiracetam could be discovered not via traditional screening methods but via its ability to block 6 Hz or audiogenic seizures. This highlights the importance of 6 Hz test that proved valuable in discovering a drug acting via a unique mechanism of action via modulation of synaptic neurotransmitter release through binding to the synaptic vesicle protein (SV2A) in the brain [3, 8, 9]. Another example is that of a novel antiseizure candidate, padsevonil, acting via binding to SV2 proteins and the $GABA_A$ receptor benzodiazepine site allowing for pre- and postsynaptic activity, that exhibited greater protection when compared to the combination of diazepam with either levetiracetam or brivaracetam in the 6 Hz test in mice [10]. Additionally, there are reports of some neurosteroids like allopregnanolone analogs exhibiting protection against focal seizures induced by 6 Hz electrical stimulation in mice possibly mediated via GABAergic receptor modulation. Thus, 6 Hz model could be an important tool in drug development for the identification of antiseizure drugs acting via GABAergic mechanisms as well [11]. It is noteworthy that the test is currently included in the Epilepsy Therapy Screening Program (ETSP) of National Institute of Neurological Disorders and Stroke (NINDS) for identifying new antiseizure drugs for therapy-resistant focal seizures. Drugs that are found to be protective against 6 Hz are later tested in other refractory models of seizures.

1.1 Characteristic Features of 6 Hz Seizures and the Effect of Antiseizure Drugs

In this model, mice show a brief period of stun position followed by minimal clonic phase, stereotyped automatism-like behavior, and other features including twitching of the vibrissae/whiskers, rearing, grooming, behavioral arrest, forelimb clonus, Straub's tail, and an increased locomotor activity, which is recorded as distance travelled over a period of 120 s. The stereotyped automatistic behavior been correlated with aura seen in patients suffering from focal or limbic seizures [2, 3]. The animals are considered to be protected if such a behavior is not observed after the shock treatment or animals resume their normal exploratory behavior within 10 s after stimulation [12]. For example, sodium valproate reduced the behavioral changes and increased the velocity (cm/s) in the 6 Hz stimulation which shows its beneficial effect [13]. The scoring of seizure, which is otherwise not used routinely, can also be utilized as an added or supplementary measure to score the efficacy of any newer antiseizure drug [14].

There are several similarities and dissimilarities between MES and 6 Hz electrical stimulation. Both models can produce electrically induced acute seizures in rodents, though MES produces GTCS and 6 Hz model generates focal seizures. The variation in MES and 6 Hz model lies in the difference between frequency and duration of electrical stimulus (Table 1). In MES, a high-frequency (60 Hz), short-duration (0.2 s), and a suprathreshold electrical stimulus is used to induce a tonic hind limb extensor response [1]. In 6 Hz, however, electrical stimulation by low frequency (6 Hz) delivered through corneal electrodes for a long duration (3 s) is employed to induce "psychomotor" seizures involving forelimb clonus and stereotyped behaviors [2]. The characteristic feature of 6 Hz paradigm is the induction of negligibly convulsive or non-convulsive seizures with automatisms, resembling "psychomotor or limbic seizures" in humans [2, 15, 16]. Occasionally, mice develop GTCS with the 6 Hz procedure, whereas with MES, tonic seizures are predictably observed [17].

Table 1
Comparison of 6-hertz and maximal electroshock seizures in mouse

	6-hertz model	Maximal electroshock model
Frequency of electrical stimulus	Low frequency (6 Hz)	High frequency (50–60 Hz)
Duration	Long duration (3 s)	Short duration (0.2 s)
Current employed	32 and 44 mA	50 mA
Electrodes used	Corneal electrodes	Corneal or ear electrodes
Type of seizures/ clinical correlate	Focal seizures resembling psychomotor or limbic seizures in humans	Generalized tonic-clonic seizures
Characteristic stages in animal	Stereotyped behavior with automatisms, stun position, twitching of vibrissae, forelimb clonus, Straub's tail, grooming, rearing, etc.	Tonic flexion, tonic hind limb extension, myoclonic jerks, stupor position, salivation, lacrimation, etc.
Useful for screening drugs for pharmacoresistant seizures	Yes	No

Numerous studies have been done to find out the influence of inducing current at different intensities on seizures and the effect of antiseizure drugs on the same in mice. The initial observations by Brown and co-workers [2] revealed that the CC_{50} of mouse psychomotor testing was 8 mA and that at 4 times CC_{50} (which was 32 mA), several antiseizure drugs available at that time (phenobarbital, trimethadione, phenacemide, etc.) show protection against 6 Hz except phenytoin, thus concluding that the model is not an effective screen for testing drugs against psychomotor seizures [2]. Barton and co-workers [3] conducted initial pharmacological studies at a stimulus intensity of 32 mA, which is equivalent to the current employed previously by Brown and co-workers [2, 3]. They also performed subsequent experiments to find out the current required to produce a seizure in 50% (CC_{50}) and 97% (CC_{97}) of the population by probit analysis. Their results suggested 19.4 mA and 22 mA as CC_{50} and CC_{97}, respectively, in mice. At 22 mA, the model could not differentiate between clinical classes of the drugs tested. However, when the current intensity was increased by 50% (32 mA), the sensitivity of certain drugs like phenytoin, carbamazepine, and lamotrigine was reduced, while topiramate was ineffective, but various other drugs like clonazepam, trimethadione, tiagabine, felbamate, ethosuximide, levetiracetam, and phenobarbital were still effective and displayed dose-dependent protection. Further, when the current intensity was doubled ($2 \times CC_{97}$ (i.e., 44 mA)), only levetiracetam and valproic acid showed full protection against 6 Hz seizure [3], and recently some novel antiseizure drugs such as brivaracetam and padsevonil demonstrated protection [10]. Hence, an interesting observation was made about the ability of levetiracetam and valproic acid to provide complete protection when all other drugs fail in the 6 Hz test at 44 mA, thereby recommending the model as a screening tool for drugs effective against refractory limbic seizures [3]. Table 2 summarizes the effect of various antiseizure drugs on the 6 Hz test at variable current intensities and their efficacy or resistance toward 6 Hz seizures. Generally, 6 Hz seizures showed resistance to antiseizure drugs acting via blockade of sodium channels; however, this was not consistent in some of the later studies.

Though the 6 Hz model has been used for several years, it is only recently that the model was characterized in the rat. The typical features of behavioral seizures induced in rats were similar to those observed in mouse. Further, the seizures correlated well with electroencephalographic (EEG) changes though the EEG changes persisted for a longer period of time even after the behavioral seizures stopped. The CC_{50} and CC_{97} observed in Sprague Dawley rats were comparable to the mouse model. However, in comparison to the mouse model, majority of the sodium channel blockers, except rufinamide, were found inactive or less efficacious in the rat 6 Hz test. The antiseizure drugs found effective at the 1.5

Table 2
Effect of antiseizure drugs on 6 Hz seizures at various current intensities in rodents

6 Hz current employed for 3 s duration	Species used	Antiseizure drugs showing protection	Antiseizure drugs showing resistance or partial resistance	References
22 mA (CC97)	Mice	Ethosuximide Lamotrigine Levetiracetam Phenytoin Valproic acid	–	[3]
32 mA (1.5 times CC97)	Mice	Clonazepam Ethosuximide Felbamate Levetiracetam Phenobarbital Tiagabine Trimethadione Valproic acid	Carbamazepine Phenytoin Lamotrigine Topiramate	[3]
	Rats	Clobazam Ethosuximide Ezogabine Levetiracetam Phenobarbitone Rufinamide Valproic acid	Carbamazepine Clonazepam Gabapentin Lacosamide Lamotrigine Phenytoin Tiagabine	[4]
44 mA (2 times CC97)	Mice	Brivaracetam Padsevonil Levetiracetam Valproic acid	Diazepam	[3, 10]
	Rats	Ezogabine Phenobarbital Valproic acid	Carbamazepine Clobazam Clonazepam Ethosuximide Gabapentin Lacosamide Lamotrigine Phenytoin Tiagabine	[4]

times CC_{97} stimulus intensity included clobazam, ethosuximide, ezogabine, levetiracetam, phenobarbital, rufinamide, and sodium valproate, while those observed to be effective at twice the CC_{97} included ezogabine, phenobarbital, and sodium valproate [4]. Interestingly, levetiracetam could not protect against 6 Hz seizures at 44 mA in the rats and required much higher ED_{50}. Similar differences were observed with other antiseizure drugs, and 44 mA stimulus intensity was considered appropriate to distinguish drugs effective in refractory cases [4]. An advantage of the rat model possibly is to compare the results with other chronic refractory models mostly carried out in rats [4].

In an attempt to validate the promising anticonvulsant compounds with little or unknown blood-brain barrier permeability, Walrave et al. (2015) reported suppression of 6 Hz-evoked seizure severity with intracerebroventricular (i.c.v.) administration of levetiracetam. In their study, seizures were induced by single application of a current intensity of 49 mA to i.c.v.-implanted NMRI mice [18].

The use of high-fat and low-carb ketogenic diet to treat epilepsy (especially pharmacoresistant seizures) is well-documented [15, 19]. Hence, the influence of ketogenic diet on 6 Hz seizure test has also been studied in mice by various researchers [15, 20, 21]. The ketogenic diet has been found to be highly sensitive by significantly elevating the seizure threshold in mice, in a time-specific manner. The CC_{50} was found to be 50.6 mA in the mice fed for 12 days with ketogenic diet and 15 mA in the mice maintained on a normal diet. However, the seizure protection was not obtained after 3 weeks' diet period suggesting the development of tolerance or compensatory mechanisms [15]. Another study reported less severe seizures and a remarkable prolongation of the electrographic response in mice receiving the ketogenic diet from the second session of 6 Hz corneal stimulation and onward [22]. Hence, it is interesting to note that the ketogenic diet has a different profile of activity from commonly used antiseizure drugs in this model, suggesting that it may act in a mechanistically distinct way from these medications [15]. It was also observed that the 6 Hz threshold was not affected by body weight in animals fed a normal diet, supporting the view that this model is applicable in conditions where there are differences in these factors between treatment groups.

1.2 6 Hz Corneal Kindling in Mice

Kindling has been a useful model for chronic epilepsy research. In addition to the various electrical and chemical kindling models described in Chapter 6 and a model of lamotrigine-resistant corneal kindled mice discussed in Chapter 10, repeated 6 Hz corneal stimulation has also shown to develop kindling in male NMRI mice. The kindling was induced by twice-daily 6 Hz stimulation via corneal electrodes using a current intensity of 44 mA for 3 s duration for 3 consecutive weeks [23]. The effect of four antiseizure drugs, viz., clonazepam, levetiracetam, valproate, and carbamazepine, on the 6 Hz was compared with the 50 Hz corneal kindling model. The antiseizure drugs tested exhibited a lower potency in 6 Hz kindling, while two drugs (levetiracetam and carbamazepine) also showed limited efficacy. Thus, the authors considered 6 Hz to be more advantageous than the established 50 Hz corneal kindling model [23].

On the other hand, Albertini et al. (2018) suggested that 6 Hz corneal kindling model in NMRI mice can also be used to study several neurobehavioral comorbidities that affect the persons with epilepsy [24]. A modified 6 Hz corneal stimulation model was recently described in which mice received up to four sessions at 32 mA with inter-stimulation interval of 72 h resulting in more severe tonic-clonic seizures after second session. Seizures resulted in EEG changes in the frontal cortex and in later stimulations involved the hippocampus. Thus, the modified method could prove useful in studying mechanisms involved in epileptogenesis [25].

1.3 Strain Differences in 6 Hz Mouse Model

It has also been stipulated that genetic makeup of mice impacts treatment resistance in 6 Hz seizure model. Leclercg and Kaminski (2015) compared seizure thresholds in the 6 Hz model in three strains of male mice (CF-1, NMRI, and C57Bl/6J) using 32 and 44 mA current intensities [8]. They observed that CF-1 mice had the lowest seizure threshold and NMRI and C57Bl/6J mice had nearly identical threshold values against phenytoin. On the other hand, levetiracetam in NMRI mice showed much higher potency and partial efficacy in CF-1 mice and low potency for C57Bl/6J mice, particularly at 44 mA. Hence, their observations clearly indicate that the selective response to a given treatment is due to genetic differences in humans and mice [8].

1.4 Ultrastructural Changes and Brain Areas Activated by 6 Hz

The ultrastructural changes in the cortex and hippocampus of mice following 6 Hz psychomotor seizures were observed using transmission electron microscopy (TEM). These changes were characterized by disorganized cytoplasm with a swollen nucleus and chromatin loss in cortical cells and disorganized mitochondria in the hippocampus. The study provided evidence for neurodegenerative changes following 6 Hz seizures. Sodium valproate effectively inhibited such changes at the ultrastructural level [26, 27].

To define the brain structures activated by 6 Hz corneal stimulation, Barton et al. [3] used the immediate early gene c-Fos as a marker for seizure-induced neuronal activation. The c-Fos immunohistochemistry studies also depicted distinct differences between 6 Hz corneal stimulation, MES-induced tonic extension, and i.v. PTZ-induced clonic convulsions. The results demonstrated widespread c-Fos induction in the dentate gyrus following MES and PTZ seizures. However, at current intensities of 22 and 32 mA in 6 Hz, the c-Fos induction remained localized to the amygdala and piriform cortex. Intense c-Fos induction in the dentate gyrus was additionally observed at 44 mA stimulus. Thus, it was discernible that at 44 mA, the apparent increased involvement of the dentate gyrus may be responsible for reduced efficacy of AEDs like levetiracetam and valproate [3].

1.5 Effect of 6 Hz Model on Neurotransmitter Levels in the Brain

It is well-known that monoamine neurotransmitters play a crucial role in the genesis and spread of epilepsy [28–31] and are altered following electroshock seizures in various animal models of epilepsy [32–34]. Since the excitatory glutamatergic neurotransmission is known to be involved in the initiation and spread of seizure activity and even a minor disinhibition in GABA can trigger hyperexcitability, a dysfunction in either GABA or glutamate availability therefore has important consequences on seizure genesis. The experiments performed in the Neurobehavioral Pharmacology Laboratory of Vohora observed reduced levels of brain serotonin, dopamine, norepinephrine, histamine, and γ-aminobutyric acid (GABA) and increased glutamate following 6 Hz psychomotor seizures in Swiss albino mice. Further, it was demonstrated that serotonin depletion by parachlorophenylalanine (PCPA) [13, 26] and noradrenaline depletion by N-2-chloroethyl-N-ethyl-2-bromobenzylamine hydrochloride (DSP-4) [27] enhanced the susceptibility of Swiss albino mice to 6 Hz psychomotor seizures. Both pre- and post-treatment with sodium valproate modulated the neurotransmitter levels (in the whole brain, cortex, and hippocampus) that accounted for its protective effects in this model [13, 26, 27].

2 Materials

Before initiating any studies, the experimental protocol of the experiment must be carried out in accordance with the in-house guidelines for care and use of laboratory animals and must be duly approved from the Institutional Animal Ethics Committee.

2.1 Animals

Mouse: To examine the antiseizure potential of newer and established drugs in this model, a variety of strains including Swiss albino, NMRI, CF-1, CD-1, C57BL/6, etc. have been used [3, 8, 13, 21, 35]. The mice were aged 4–6 weeks and 25–35 g in weight. The standard facilities of animal house like controlled temperature at 20 °C and 50–55% humidity and 12 h day and night cycle, with free access to food and water, are followed.

Rats: This model is usually performed in the mouse, but few researchers have tried to duplicate the model in rats [4, 35, 36]. As discussed in Subheading 1.2, Metcalf and co-workers have characterized the model in Sprague Dawley rats and concluded that the same may be employed successfully for screening drugs for pharmacoresistant seizures [4].

Other researchers have compared the 6 Hz model in rats and mice and observed that though the magnitude of seizures was different in the two animal species, the antiseizure drugs (phenytoin, lamotrigine, levetiracetam, valproate) show similar efficacy [35].

2.2 Electro-convulsometer with Corneal Electrodes

Electroconvulsometer (ECT) unit, UgoBasile57800-001, Italy, or any suitable ECT may be used for inducing seizures (see point number 2, notes section 4). The UgoBasile unit supplies from 1 to 299 rectangular pulses per second, 0.1 to 0.9 ms in width, for duration variable from 0.1 to 9.9 s. Shock current can be accurately pre-set on a scale of 1–99 mA.

2.3 Videopath Activity Analyzer or a Transparent Plexiglass Chamber with Camera

Videopath activity analyzer (Coulbourne, USA) or any other suitable activity monitor may be used for recording the movement of the animal in the chamber (distance travelled, velocity cm/s). Observations will also be made manually by recording the onset, number of episodes and total duration of Straub's tail, stun position, forelimb clonus, twitching of vibrissae, rearing, grooming, etc.

3 Methodology

The animals are kept in a transparent Plexiglas chamber ($24 \times 17 \times 13$ cm) in a Videopath activity analyzer with a camera set on the vertical stand above it to record the epileptic behaviors in mice following 6 Hz stimulation [13, 26, 27]. Free access to standard pellet diet and water is provided, except during testing period. Wood chips or powder is to be used in all cages. Experiments are to be performed during the light phase of the natural light/dark cycle after a minimum 30 min period of acclimation to the experimental room. Mice are then randomly divided into groups as control and drug-treated. The drug treatment is followed for specified days, following which the mice are subjected to 6 Hz stimulation [13, 37]. The current required to produce seizure in 50% (CC_{50}) and 97% (CC_{97}) of the animal population is calculated by probit analysis [3]. Each animal is employed only once during the study. The test drug's ability to prevent seizures induced by 6 Hz corneal stimulation is evaluated. In mice, a current intensity of 22 mA is adequate to evoke a seizure in 97% of the population tested (CC_{97}) [3]; however, this may be standardized as per one's laboratory conditions. The 32 mA current stimulation is 1.5 times the CC_{97}, and the 44 mA stimulation is 2 times the CC_{97}, as discussed earlier [3, 13].

3.1 Experimental Procedure

1. Test drug treatment and standard antiseizure drugs may be administered for a specified duration.

2. A drop of topical anesthetic (0.5% tetracaine hydrochloride ophthalmic solution) is applied to the eyes of the mice or rats, 30 min prior to the experimental process. Later, a drop of 0.9% saline is administered to the eyes of mice or rats, preceding the position of corneal electrode to confirm good electrical contact.

3. Seizure is induced by corneal stimulation (6 Hz, 0.2 ms rectangular pulse width, 3 s duration) using electroconvulsometer/stimulator (ECT unit, UgoBasile57800-001, Italy). Current administer is 32 mA for simple psychomotor seizures or 44 mA for evaluating refractory seizures. This may be selected based on the objectives of the study (also see point number 3, notes section 4).

4. Animal is manually controlled and discharged quickly after taking the 6 Hz stimulation and observed for the presence/absence of seizure activity and behavior in a Videopath activity analyzer connected to a camera.

5. The total cumulative duration of behavioral seizures characterized as stun position, twitching of the vibrissae, Straub's tail, forelimb clonus, and other behavioral effects including rearing, grooming, and distance travelled (velocity) is recorded in a transparent chamber/Videopath activity analyzer for 120 s (see point number 1, notes section 4).

6. If any of the stereotyped behavior mentioned above is not displayed and if they resume their normal exploratory behavior within 10 s of 6 Hz stimulation, the animals are considered as protected.

7. The effect of an antiseizure drug is quantitated by determining % protection at various doses or by statistically evaluating the reduction of duration of each seizure stage.

4   Notes

1. Stun position is one of the characteristic features of 6 Hz seizures. However, a mouse may remain immobile in stun position despite not sustaining a seizure, and hence it is always better to pinch the tail with a clamp that exerts 70 g pressure, approximately 1 in. from the body, to check if it vocalizes and withdraws. The animal experiencing seizure is not affected by this stimulus [2].

2. While operating the ECT unit, the experimenter must always wear rubber gloves when handling electrodes connected to the instrument in a live experiment.

3. The 32 and 44 mA currents, though widely employed in mice, must be standardized under your own laboratory conditions, and the effect of antiseizure drugs at various doses/duration (levetiracetam and others) must be administered to confirm their efficacy or resistance to a particular stimulus/current to determine at what dose and duration it indeed reflects refractory seizures.

4. Once the efficacy of the test drug is confirmed in 6 Hz, the study may be continued further to investigate its effect on other known models of refractory chronic epilepsy.

5 Limitations of 6 Hz Model

Despite offering various advantages over other models of seizures, the 6 Hz model has few drawbacks. Being a model of acute seizure, it does not correlate the neuronal and pathophysiological adaptations occurring in the brain during chronic epilepsy. In humans and in chronic models of epilepsy, variations in remodeling of neuronal circuits, role of inflammatory mediators, and the modifications in protein expressions have been reported [7]. These variations occurring in neurons after chronic diseased condition cannot be studied in acute models like 6 Hz.

The analysis of c-Fos immunoreactivity in 6 Hz model shows the involvement of different regions of the brain like neocortex, cingulate, piriform cortex, amygdala, caudate putamen, etc., but the data proving the involvement of the hippocampus is lacking [3, 38]. As the hippocampus has been frequently shown to be involved in refractory seizures, this is considered as a drawback of 6 Hz model (also see point number 4 of Notes section). However, Giordiano et al. (2015) have reported the late involvement of the hippocampus in the seizure spread in the repeated 6 Hz stimulation model [25].

Further, unlike other refractory seizure models, 6 Hz model does not distinguish between responders and non-responders to antiseizure drug, there are no spontaneous recurrent seizures, and this is not useful for studying the mechanisms of drug resistance [3, 4].

6 Conclusion

A considerable number of persons with epilepsy are refractory to current antiseizure treatment regimen regardless of the availability of the newer and traditional drugs. Hence, there is a necessity for more effective antiseizure drugs for such persons. The process of finding newer antiseizure drug relies heavily on acute in vivo seizure models like MES, s.c. PTZ, kindling models, and 6 Hz model in rodents. The precise mechanisms that contribute to pharmacoresistance and how few drugs like levetiracetam, padsevonil, allopregnanolone analogs, etc. are effective in reversing this state are also unclear. Hence, there is still a need of more advanced scientific investigations in the field of resistant seizures in a simple model such as 6 Hz. Though some newer antiseizure drugs showed efficacy in the 6 Hz model, the model seems to be particularly refractory to sodium channel blockers and hence may not be considered as a model for pharmacoresistant seizures in general. Nevertheless, the model is a useful tool to differentiate drugs while screening for antiseizure activity in a drug discovery and development program.

References

1. Swinyard EA (1972) Electrically induced seizures. In: Purpura DP, Penry JK, Tower DB, Woodbury DM, Walter RD (eds) Experimental models of epilepsy: a manual for the laboratory worker. Raven Press, New York, pp 433–458

2. Brown WC, Schiffman DO, Swinyard EA, Goodman LS (1953) Comparative assay of antiepileptic drugs by "psychomotor" seizure test and minimal electroshock threshold test. J Pharmacol Exp Ther 107(3):273–283

3. Barton ME, Klein BD, Wolf HH, White HS (2001) Pharmacological characterization of the 6Hz psychomotor seizure model of partial epilepsy. Epilepsy Res 47(3):217–227

4. Metcalf CS, West PJ, Thomson KE, Edwards SF, Smith MD, White HS, Wilcox KS (2017) Development and pharmacologic characterization of the rat 6 Hz model of partial seizures. Epilepsia 58(6):1073–1084

5. White HS, Bender AS, Swinyard EA (1988) Effect of the selective N-methyl-D-aspartate receptor agonist 3-(2-carboxypiperazin-4-yl) propyl-1-phosphonic acid on [3H] flunitrazepam binding. Eur J Pharmacol 147(1):149–151

6. Loscher W, Honack D (1993) Profile of UCB L059, a novel anticonvulsant drug, in models of partial and generalized epilepsy in mice and rats. Eur J Pharmacol 232:147–158

7. Klitgaard H, Matagne A, Gobert J, Wulfert E (1996) Levetiracetam (UCB L059) prevents limbic seizures induced by pilocarpine and kainic acid in rats. Epilepsia 37(S5):118

8. Leclercq K, Kaminski RM (2015) Status epilepticus induction has prolonged effects on the efficacy of antiepileptic drugs in the 6-Hz seizure model. Epilepsy Behav 49:55–60

9. Gower AJ, Noyer M, Verloes R, Gobert J, Wülfert E (1992) UCB L059, a novel anticonvulsant drug: pharmacological profile in animals. Eur J Pharmacol 222(2–3):193–203

10. Leclercq K, Matagne A, Provins L, Klitgaard H, Kaminski RM (2020) Pharmacological profile of the novel antiepileptic drug candidate padsevonil: characterization in rodent seizure and epilepsy models. J Pharmacol Exp Ther 372(1):11–20

11. Kaminski RM, Livingood MR, Rogawski MA (2004) Allopregnanolone analogs that positively modulate GABAA receptors protect against partial seizures induced by 6-Hz electrical stimulation in mice. Epilepsia 45(7):864–867

12. Wojda E, Wlaz A, Patsalos PN, Luszczki JJ (2009) Isobolographic characterization of interactions of levetiracetam with the various antiepileptic drugs in the mouse 6 Hz psychomotor seizure model. Epilepsy Res 86(2–3):163–174

13. Jahan K, Pillai KK, Vohora D (2017) Parachlorophenylalanine-induced 5-HT depletion alters behavioral and brain neurotransmitters levels in 6-Hz psychomotor seizure model in mice. Fundam Clin Pharmacol 31(4):403–410

14. Racine RJ (1972) Modification of seizure activity by electrical stimulation: II. Motor seizure. Electroencephalogr Clin Neurophysiol 32(3):281–294

15. Hartman AL, Lyle M, Rogawski MA, Gasior M (2008) Efficacy of the ketogenic diet in the 6-Hz seizure test. Epilepsia 49(2):334–339

16. Bankstahl M, Bankstahl JP, Löscher W (2013) Pilocarpine-induced epilepsy in mice alters seizure thresholds and the efficacy of antiepileptic drugs in the 6-hertz psychomotor seizure model. Epilepsy Res 107(3):205–216

17. Rowley NM, White HS (2010) Comparative anticonvulsant efficacy in the corneal kindled mouse model of partial epilepsy: correlation with other seizure and epilepsy models. Epilepsy Res 92(2–3):163–169

18. Walrave L, Maes K, Coppens J, Bentea E, Van Eeckhaut A, Massie A, Van Liefferinge J, Smolders I (2015) Validation of the 6 Hz refractory seizure mouse model for intracerebroventricularly administered compounds. Epilepsy Res 115:67–72

19. Freeman J, Veggiotti P, Lanzi G, Tagliabue A, Perucca E (2006) The ketogenic diet: from molecular mechanisms to clinical effects. Epilepsy Res 68(2):145–180

20. Neal EG, Chaffe H, Schwartz RH, Lawson MS, Edwards N, Fitzsimmons G, Whitney A,

Cross JH (2008) The ketogenic diet for the treatment of childhood epilepsy: a randomised controlled trial. Lancet Neurol 7(6):500–506

21. Giordano C, Marchiò M, Timofeeva E, Biagini G (2014) Neuroactive peptides as putative mediators of antiepileptic ketogenic diets. Front Neurol 29(5):63

22. Lucchi C, Marchiò M, Caramaschi E, Giordano C, Giordano R, Guerra A, Biagini G (2017) Electrographic changes accompanying recurrent seizures under ketogenic diet treatment. Pharmaceuticals 10(4):82

23. Leclercq K, Matagne A, Kaminski RM (2014) Low potency and limited efficacy of antiepileptic drugs in the mouse 6 Hz corneal kindling model. Epilepsy Res 108(4):675–683

24. Albertini G, Walrave L, Demuyser T, Massie A, De Bundel D, Smolders I (2018) 6Hz corneal kindling in mice triggers neurobehavioral comorbidities accompanied by relevant changes in c-Fos immunoreactivity throughout the brain. Epilepsia 59(1):67–78

25. Giordano C, Vinet J, Curia G, Biagini G (2015) Repeated 6-Hz corneal stimulation progressively increases FosB/ΔFosB levels in the lateral amygdala and induces seizure generalization to the hippocampus. PLoS One 10 (11):e0141221

26. Jahan K, Pillai KK, Vohora D (2019) Serotonergic mechanisms in the 6-Hz psychomotor seizures in mice. Hum Exp Toxicol 38 (3):336–346

27. Jahan K, Pillai KK, Vohora D (2018) DSP-4 induced depletion of brain noradrenaline and increased 6-hertz psychomotor seizure susceptibility in mice is prevented by sodium valproate. Brain Res Bull 142:263–269

28. Vohora D, Pal SN, Pillai KK (2001) Histamine and selective H3-receptor ligands: a possible role in the mechanism and management of epilepsy. Pharmacol Biochem Behav 68 (4):735–741

29. Fujimoto Y, Funao T, Suehiro K, Takahashi R, Mori T, Nishikawa K (2015) Brain serotonin content regulates the manifestation of tramadol-induced seizures in rats: disparity between tramadol-induced seizure and serotonin syndrome. Anesthesiology 122 (1):178–189

30. Cumper SK, Ahle GM, Liebman LS, Kellner CH (2014) Electroconvulsive therapy (ECT) in Parkinson's disease: ECS and dopamine enhancement. J ECT 30(2):122–124

31. Barton ME, Peters SC, Shannon HE (2003) Comparison of the effect of glutamate receptor modulators in the 6 Hz and maximal electroshock seizure models. Epilepsy Res 56 (1):17–26

32. Nakamura J, Mine K, Yamada S (1991) Effects of anticonvulsants on the electroconvulsive threshold lowered by DA, 5-HT or GABA depletion. Kurume Med J 37(4):253–259

33. Scherkl R, Hashem A, Frey HH (1991) Histamine in brain—its role in regulation of seizure susceptibility. Epilepsy Res 10(2–3):111–118

34. Jobe PC, Stull RE, Geiger PF (1974) The relative significance of norepinephrine, dopamine and 5-hydroxytryptamine in electroshock seizure in the rat. Neuropharmacology 13 (10–11):961–968

35. Esneault E, Peyon G, Castagné V (2017) Efficacy of anticonvulsant substances in the 6 Hz seizure test: comparison of two rodent species. Epilepsy Res 134:9–15

36. Metcalf CS, Huff J, Thomson KE, Johnson K, Edwards SF, Wilcox KS (2019) Evaluation of antiseizure drug efficacy and tolerability in the rat lamotrigine-resistant amygdala kindling model. Epilepsia Open 4(3):452–463

37. Tomaciello F, Leclercq K, Kaminski RM (2016) Resveratrol lacks protective activity against acute seizures in mouse models. Neurosci Lett 632:199–203

38. Duncan GE, Kohn H (2005) The novel antiepileptic drug lacosamide blocks behavioral and brain metabolic manifestations of seizure activity in the 6 Hz psychomotor seizure model. Epilepsy Res 67(1–2):81–87

39. Klein P, Diaz A, Gasalla T, Whitesides J (2018) A review of the pharmacology and clinical efficacy of brivaracetam. Clin Pharmacol 10:1–22

Chapter 10

A Method to Induce Lamotrigine-Resistant Corneal Kindled Mice

Melissa Barker-Haliski

Abstract

The historical identification of promising investigational therapies for the treatment of epilepsy can be credited to the use of several animal models of seizure and epilepsy for over 80 years. In this time, three core preclinical models have advanced all approved antiseizure medicines (ASMs): the maximal electroshock (MES) test in mice and rats, the subcutaneous pentylenetetrazol (scPTZ) test in mice and rats, and rat kindling models. Kindling models in rodents in general are particularly useful to ASM discovery because they lead to chronic network hyperexcitability and behavioral changes consistent with human temporal lobe epilepsy. However, traditional use of focal kindling models in rats has been limited to late-stage differentiation tests, rather than moderate- to high-throughput drug screening applications, because of the complex surgical procedures needed to implant stimulating and recording electrodes and the large size of rats that prohibits the early evaluation of investigational compound libraries. Furthermore, despite the more than 20 ASMs on the market today, nearly 30% of patients with epilepsy remain resistant to these currently available medications. To address this unmet medical need, the early identification of novel agents for the treatment of epilepsy must now increasingly employ models of pharmacoresistant seizures in organisms that are appropriately sized to accommodate early, moderate- to high-throughput drug screening. Drug-resistant kindled rats emerged in the 1990s and early 2000s as a suitable platform for ASM differentiation. However, we have more recently developed the lamotrigine-resistant corneal kindled mouse model as a uniquely positioned pharmacoresistant chronic seizure model that is suitable to moderate- to high-throughput ASM discovery. This model can reliably produce large numbers of drug-resistant kindled mice in a relatively short time period for subsequent moderate-throughput ASM screening. The lamotrigine-resistant kindled mouse demonstrates robust resistance to other ASMs including retigabine, valproic acid, and other sodium channel-blocking ASMs like phenytoin and carbamazepine. The goal of this present text is to thus provide the reader with guidelines to similarly generate this lamotrigine-resistant corneal kindled mouse model that is well-suited for early ASM discovery.

Key words Antiseizure medications, Lamotrigine, Corneal kindling, Mouse, Epilepsy, Drug screening, Drug-resistant epilepsy

Divya Vohora (ed.), *Experimental and Translational Methods to Screen Drugs Effective Against Seizures and Epilepsy*, Neuromethods, vol. 167, https://doi.org/10.1007/978-1-0716-1254-5_10,
© Springer Science+Business Media, LLC, part of Springer Nature 2021

1   Introduction: Kindling Models of Chronic Network Hyperexcitability and Utility of Corneal Kindling Models for Drug Discovery

Although numerous antiseizure medicines (ASMs) are approved by the US Food and Drug Administration (FDA) and European Medicines Agency (EMA) and listed on the World Health Organization's essential medicines list (https://www.who.int/medicines/publications/essentialmedicines/en/), a significant proportion of individuals with epilepsy remain refractory to therapy and are thus considered drug resistant [1]. Over 65 million people live with epilepsy worldwide, leaving a significant unmet therapeutic need to continue to identify and develop new and improved treatments for this pharmacoresistant epilepsy patient group. Basic science has made incredible strides in understanding the pathophysiology of epilepsy, including drug-resistant epilepsy [2, 3]. Further, we have developed in recent years numerous preclinical models of drug-resistant epilepsy. The challenge is now to translate the identification of promising compounds into transformative therapies for those patients who are still refractory to currently available therapies, as well as to identify therapies that may have better safety and tolerability profiles to improve patient adherence to prescribed medications [4] and reduce adverse effect liabilities [5], all of which may also negatively influence uncontrolled or pharmacoresistant epilepsy.

The precise mechanisms that contribute to pharmacoresistant epilepsy remain unclear and subject of vigorous scientific investigation [2, 6]. Pharmacoresistant epilepsy itself contributes to greater economic burden [7], more severe comorbidities [8], and increased risk of sudden unexplained death in epilepsy (SUDEP; [9]). Despite continuous advancements in our understanding of the underlying disease biology and many new ASMs, the percentage of patients with epilepsy who do not achieve seizure freedom with available medications has remained unchanged for decades [1]. There is thus a great need for more effective pharmacotherapies for drug-refractory epilepsy.

The traditional approach to ASM discovery has relied primarily on the evaluation of acute anticonvulsant efficacy in two clinically validated models of acute electrically or chemically evoked seizures in naïve rodents [3, 10, 11]: the maximal electroshock (MES) test and the s.c. pentylenetetrazol test (scPTZ). MES is a model of generalized tonic-clonic seizures and provides an indication of a compound's ability to prevent seizure spread. The MES test was validated with the initial identification of phenytoin by Merrit and Putnam in 1937; 1 year later this agent was available for clinical use [12]. MES-induced seizures are highly reproducible and are electrophysiologically consistent with human seizures. The scPTZ test is generally considered a useful model of generalized non-convulsive myoclonic and generalized spike-wave seizures [13, 14]. Validation of the scPTZ test was provided largely by the findings that trimethadione was effective against PTZ-induced seizures [15]. Lennox subsequently demonstrated that trimethadione was effective against absence seizures, but was ineffective or worsened generalized tonic-clonic seizures [16]. Trimethadione's clinical success and its ability to block PTZ-induced threshold seizures provided sufficient evidence to establish the scPTZ test as a model of generalized absence seizures. The scPTZ model is thus useful to identify compounds that may be effective in patients with primary generalized spike-wave and myoclonic seizures. Due to their clinical validation, technical feasibility, and potential to differentiate activity of ASMs, the MES and scPTZ tests are heavily utilized for early ASM identification in most screening programs [11].

Seizures are just one symptom of the neurological disorder of epilepsy. Unfortunately, the acute MES and scPTZ seizure models altogether lack the behavioral and pathophysiological changes associated with chronic seizures and epilepsy, e.g., network remodeling, neuroinflammation, and alterations in protein expression. Thus the early evaluation of potential ASMs has also included differentiation studies in etiologically relevant rodent models that are defined by chronic seizures and behavioral deficits [17]. Rodents with chronic and/or spontaneous seizures are generally available in the drug discovery armamentarium [18], but most of these models are incredibly resource- and labor-intensive and are therefore not suitable for frontline identification screening [3, 19]. As also discussed in Chapter 6, rodent kindling produces a model of chronic network hyperexcitability and seizures that has been historically used for ASM discovery for several decades [20, 21]. In fact, kindling in rats was validated with the identification of levetiracetam [22, 23], and this model is now commonplace in ASM discovery practice. However, kindling in rats generally requires focal implantation of stimulation electrodes into the brain [24, 25], making this model quite low-throughput. Rodent chronic seizure models do not always exhibit a discernable pharmacoresistance [26, 27] and may thus not be suitable to identify treatments for drug-resistant epilepsy.

Kindling in rats was originally described by Goddard [28, 29] and then scaled down to mice using corneal stimulation by Sangdee and colleagues [30] before the pharmacology of the corneal kindled mouse was characterized by Matagne and Klitgaard over 16 years later [31]. Corneal kindled mice have been particularly beneficial to moderate-throughput ASM discovery because of the ease with which large numbers of animals with uniform seizure history can be generated in a small amount of time without significant technical requirements. Specifically, no surgical equipment is required to corneal kindle mice; the electrical stimulus is delivered through electrodes applied bilaterally to anesthetized corneas. Corneal kindled mice are now a well-established moderate-throughput drug screening model that is suitable for early ASM discovery, especially when compound supply is limited in the early phases of chemical library synthesis [26, 31].

There are numerous benefits to the use of the corneal kindled mouse for early drug discovery. First, the 50/60 Hz corneal kindled mouse demonstrates a pharmacological profile consistent with the hippocampal kindled rat model [26] but requires far less active pharmaceutical ingredient for the early evaluation of potential efficacy against drug-resistant seizures [31] due to the simple difference in body mass between rats and mice. Second, the corneal kindled mouse exhibits neuroinflammation (e.g., reactive gliosis [32, 33]) and behavioral alterations [34, 35] consistent with human TLE. Third, the pharmacological profile of the corneal kindled mouse is also consistent with human focal impaired awareness epilepsy and effectively identifies the anticonvulsant potential of useful compounds for this condition, such as levetiracetam [26, 36, 37]. Thus, novel compounds and/or chemical libraries that are not active in MES or scPTZ assays should be tested in the corneal kindled mouse to assess potential anticonvulsant activity prior to abandoning any further development of a molecule.

Kindling models mimic, at least in part, the epileptogenesis process and are widely used because of the ability of these models to predict clinically useful drugs with anticonvulsant efficacy against focal and secondarily generalized seizures [22]. However, one major limitation to traditional kindling models, whether induced in mice or rats, is that these models are not typically drug resistant. ASMs are generally effective at non-motor impairing doses in the 50/60 Hz corneal kindled mouse and hippocampal kindled rat [26, 38]. The amygdala-kindled rat also does not exhibit a marked shift in ASM efficacy [11, 37]. Despite the US NINDS Epilepsy Therapy Screening Program's effort to revise ASM discovery efforts to use the 60 Hz corneal kindled mouse model early in ASM discovery [10], this mouse model also does not demonstrate a drug-resistant seizure phenotype. In this regard, early ASM identification does not presently heavily rely on moderate-throughput models of pharmacoresistance. To address this unmet need, we developed a lamotrigine (LTG)-resistant corneal kindled mouse model that is refractory to sodium channel-blocking ASMs (e.g., carbamazepine and LTG), demonstrates a reduced sensitivity to valproic acid, and has resistance to the novel ASM, retigabine [39]. This innovative model of drug-resistant corneal kindled seizures carries significant potential to identify novel therapies that may be useful for drug-resistant epilepsy patient populations. Implementing the LTG-resistant corneal kindled mouse model into ASM discovery programs should be prioritized to identify transformative therapies that may effectively identify treatments for drug-resistant epilepsy.

2   Materials

2.1   Mouse Strain and Age

The LTG-resistant corneal kindled mouse was originally characterized in male, 3–4-week-old (12–14 g) CF-1 mice from Charles River Laboratories in the United States. Kindling rates can dramatically differ as a function of mouse sex and age alone [40, 41]. It is well-known that mouse strain can also significantly affect seizure threshold [42] and the efficacy of ASMs in various acute seizure tests [43, 44]; thus the investigator is first strongly advised to determine the minimal clonic seizure threshold of the mouse strain, age, and sex in use for any study prior to initiating a kindling protocol. The stimulation current for corneal kindling should then be empirically determined based on the minimal clonic seizure threshold of the mouse strain, age, and sex in use (discussed below) such that the stimulation current is initially benign. The current should be sufficient to ultimately induce a generalized seizure over the course of 5–7 days [40].

2.2   Minimal Clonic Seizure Threshold Testing

This step-by-step protocol for the minimal clonic seizure threshold test is based on that previously described [25, 40]. Minimal clonic seizures are characterized by rhythmic face and forelimb clonus and ventral neck flexion and may include rearing and falling but not maximal tonic hindlimb extension [45, 46], characteristic of an MES seizure [47]. Minimal clonic seizures are elicited by a 60 Hz, 0.2 s, 0–20 mA current, depending on the mouse strain and age in use [42, 40]. Briefly, an electrical current should be delivered bilaterally to anesthetized (0.5% tetracaine) corneas. A mouse is considered to have had a minimal clonic seizure if it presents with rhythmic face and forelimb clonus, ventral neck flexion, and even rearing and falling. The stimulation that elicits tonic hindlimb extension is considered to exceed that which is necessary for a minimal clonic seizure; if this seizure results, the stimulation current should be accordingly reduced. To determine the median convulsant current (CC50), the seizure result of each mouse dictates the stimulation current delivered to the next mouse in each group until at least five current groups ($n = 7$–9 mice/group is recommended) can be established between the limits of 0% and 100% of mice with seizure to quantify CC50 and 95% confidence intervals.

2.3   Animal Husbandry

Animals should be housed on a fixed light: dark cycle within an established and approved vivarium. Mice are to be given free access to food (with sterilization protocol noted, if available, i.e., autoclaved or irradiated) and water (filtered or purified in a controlled manner, with method appropriately noted), except during the periods of behavioral manipulation. Mice should be group-housed with standard bedding materials and free access to enrichment materials (e.g., Nestlets, cardboard tubes, huts, etc.) because the corneal kindling model itself does not require recording electrodes and thus single-housing. Husbandry conditions should be held constant throughout the study period. All procedures should be approved by the Institutional Animal Care and Use Committee or similar governing body and conform to the *ARRIVE* guidelines [48].

For all anticonvulsant efficacy studies in kindled rodents, it is recommended that an investigator perform appropriate and similar handling of all experimental animals *before* the start of in-life testing (i.e., handling habituation) to reduce bias; acute stress mostly exerts anticonvulsant effects, whereas chronic stress conditions might worsen seizures [25, 49]. Vehicle-treated mice should receive similar handling and electrical stimulation as the LTG-kindled group. To alleviate any potential confounds of stress-induced effects on seizure susceptibility, sham-stimulation control groups should receive mock stimulations and similar handling, as well.

2.4 Corneal Kindling Apparatus

Electroconvulsant kindled seizures are induced in mice through corneal electrodes. Electrodes and the stimulus are applied bilaterally to anesthetized, open eyes. A 0.9% saline and local anesthetic (e.g., tetracaine/lidocaine) solution is applied to the eyes to improve electrical conductance and minimize irritation from the electrode placement. The time interval between local anesthesia and stimulation depends on activity of the local anesthetic. A specialized device for electrical stimulation of rodents is suitable to deliver the electrical stimulus needed for corneal kindling, as long as the instrument is capable of delivering sufficient electrical current necessary to evoke a seizure [50]. It is important to use the same stimulation device and electrical stimulation train once a study has been initiated and model characterized (e.g., the ECT (electroconvulsive therapy)) unit from Ugo Basile [51]. Lastly, electrical wave shape should also be noted (e.g., square wave) because it could also influence the behavioral seizure itself. In our original characterization of the LTG-resistant corneal kindled mouse, we used a sinusoidal pulse delivered by an electrical stimulation apparatus similar to that which was originally described by Woodbury and Davenport [52]. This electrical stimulator is capable of delivering a 60 Hz constant current electrical stimulation of varying current intensities (i.e., 3 mA for CF-1 mice and 1.6 mA for C57Bl/6 mice).

2.5 Establishing the Dose of Lamotrigine (LTG) to Be Used

The dose of LTG (AK Scientific catalogue #K499), formulated in 0.5% methylcellulose (MC; Sigma-Aldrich catalogue #M0430) used in the original description of the LTG-resistant corneal kindled mouse, was based on the median effective (ED50) dose of this drug in the mouse MES [39]. The reader is thus first urged to establish the MES ED50 for LTG in the mouse strain in use within the investigational laboratory [25]. For male CF-1 mice aged

1–2 months old, this dose of LTG has been previously reported [53]. However, if other mouse strains are to be used, the time- and dose-response evaluation of LTG in the MES test should be performed to establish an anticonvulsant dose to then administer to mice during the kindling process.

3 Methods

The kindling model is a chronic model of network hyperexcitability that results when repeated stimulation with an initially benign electrical stimulus reliably elicits evoked seizures after days to weeks of repeated electrical stimulation [28]. Although there are indeed chemoconvulsant and even optogenetic kindling models [54, 55], we will herein only focus our discussion on electrical kindling, but it is certainly reasonable to assume that LTG-resistant chemoconvulsant or optogenetic kindling models could be developed as there are both a LTG-resistant amygdala-kindled rat model [56, 57] and a LTG-resistant chemoconvulsant kindling model [58]. Regardless of the approach, the initially non-convulsive electrical stimulus induces a permanent seizure susceptibility and lasting brain alterations that are similar to those found in human temporal lobe epilepsy [59]. The pharmacology of the kindled rodent suggests that this model generally displays the best predictive validity for human focal epilepsy. However, it is important to note that at the time at which an animal is considered to be fully kindled, seizures are still evoked and not spontaneous. For drug screening, the effect of an investigational agent on the subsequent expression of the evoked secondarily generalized focal seizure is the primary outcome measure. Due to these technical considerations, the kindled rodent has not historically been considered a true model of "epilepsy" wherein the animal displays spontaneous, unprovoked seizures suitable for late-stage drug differentiation. The characterization of pharmacoresistant kindled rodents is highly useful to support preclinical differentiation and/or clinical studies [39, 56, 60]. Indeed, the kindled rodent is the only *validated* model commonly used for early ASM screening that recapitulates many comorbid features of human temporal lobe epilepsy, including anxiety-like behaviors, depression, and cognitive deficits. Corneal kindled mice are particularly more well-suited for early-phase moderate-throughput ASM discovery than the acute seizures elicited in neurologically intact rodents (e.g., MES/scPTZ) because kindling more closely reproduces human focal epilepsy. Finally, kindling models are also suited to both evaluation of antiepileptogenic agents (e.g., those that prevent the development or modify the severity of epilepsy; [61]) and to the evaluation of anticonvulsant agents (e.g., therapies that block acute symptomatic seizures [31, 62–66]).

3.1 Corneal Kindling   Commercially acquired mice should be allowed a minimum 4-day acclimatization period within the vivarium prior to being randomized to corneal or sham kindling protocols to minimize stress due to transport.

3.1.1 Corneal Kindling Groups   Cohorts of mice to be corneal kindled should be divided into LTG and vehicle-treatment groups. Group sizes should be appropriately powered for the long-term study objectives and based on the number of animals that typically achieve kindling criterion in the mouse strain, age, and sex in use. Group sizes should be sufficiently powered for downstream analysis to meet study objectives (e.g., drug screening, behavior, histopathology, etc.).

3.1.2 Pharmacological Induction of LTG Resistance

1. The corneal kindling cohort should be further divided into two treatment groups: intraperitoneal (i.p.) administration of 0.5% MC vehicle (Sigma-Aldrich, #M0430) or administration of LTG (8.5 mg/kg) 30 min before each twice-daily corneal stimulation (60 Hz, 3 s, 3 mA stimulation).

2. Twice-daily corneal stimulations continue on a Monday-Friday frequency until each mouse has achieved the criterion of five consecutive Racine stage 5 seizures, whereby it is then considered "fully kindled." In our study of CF-1 mice from Charles River, corneal kindled mice received a 3 s (60 Hz, 3 mA intensity) stimulation twice daily until acquisition of kindling criterion defined as five consecutive Racine stage 5 seizures. Stage 5 seizures were reached after twice-daily corneal stimulation for 8–10 days.

3. Twice-daily stimulations must be separated by a minimum of 4 h and continued until each mouse achieves the criterion of five consecutive Racine stage 5 seizures, whereby it is considered "fully kindled." Racine stage seizures are defined as follows: stage 1, vibrissae twitching, jaw chomping, and Straub tail; stage 2, stage 1 + head bobbing; stage 3, stage 2 + unilateral forelimb clonus; stage 4, stage 3 + bilateral forelimb clonus and rearing; and stage 5, stage 4 + rearing and falling [67]. For the corneal kindling procedure, the mouse is gently restrained in the hand of the investigator, a drop of topical electrolyte/anesthetic solution (0.9% saline/0.5% tetracaine) applied to each eye, and the stimulation electrode bilaterally applied to deliver the current to each eye (*see* **Note 1**). The mouse is then released and observed for the presence or absence of the seizure, which is scored according to the Racine scale [25, 26, 31].

4. Fully kindled mice are then stimulated every other day until all mice within the group reach the criterion of five consecutive stage 5 seizures (*see* **Note 2**).

5. Any mouse not achieving the fully kindled state by 17 days post-initiation of the twice-daily kindling protocol should not be included in any subsequent behavioral or pharmacological testing (for CF-1 strain from Charles River; *see* **Note 2**). It is advisable to remove the animal from study to minimize potential for confounding of study results.

6. For sham kindling of mice, animals should receive twice-daily handling, topical anesthetic, and electrode application without electrical stimulation. All other behavioral parameters should be held constant and consistent between corneal and sham kindling cohorts throughout the study period.

7. Testing of LTG resistance and investigational compounds in fully kindled and sham kindled mice should commence at least 5–7 days after the last corneal stimulation necessary for all mice to be fully kindled. Corneal kindled seizures are typically stable for weeks after acquisition of the fully kindled state [30].

3.1.3 Lamotrigine Challenge to Confirm LTG Resistance

Upon acquisition of the fully kindled state and a 5–7-day recovery period, mice can begin drug administration or behavioral testing, according to the particular study objectives. All drug testing sessions require a baseline stimulation session 24 h prior to the drug testing day to confirm the baseline seizure severity for comparison on the drug testing day. Only mice that present with a Racine stage 5 seizure on the baseline stimulation session should be included for subsequent drug evaluation studies.

1. Prior to beginning any further behavioral testing, mice should be challenged with an increased dose of LTG to confirm LTG resistance (or sensitivity for vehicle-kindled animals). In our study in CF-1 mice, we administered 17 mg/kg LTG, which was a twofold increase from that which was administered repeatedly during the kindling procedure, i.e., 8.5 mg/kg, i.p. [39]. Mice should receive this increased dose of LTG on the testing day following a 1-h habituation period in the testing room. One hour later, the same stimulation current is delivered through corneal electrodes and the seizure severity scored. Seizure scores of ≤ 2 are considered "protected." The mean seizure score ($\pm$ S.E.M.) for the LTG versus VEH treatment group should be compared by Student's *t*-test (or ANOVA if multiple LTG-treated kindling groups are used) to determine statistical significance.

2. Following the determination of LTG resistance, mice that are resistant to LTG should be used for subsequent behavioral testing throughout the study period. Fully kindled mice are used for investigational drug screening a minimum of 24–48 h later to define their sensitivity to other ASMs and/or mechanistic compounds over the following several weeks to months,

depending on the study objectives. Corneal kindled mice, whether LTG-resistant or not, can be used repeatedly for ASM discovery and screening for up to 4 months post-kindling acquisition with appropriate washout between drug administration sessions to minimize potential for drug-drug interactions.

3.1.4 Health Assessment During Kindling Acquisition

It is necessary to track whether there are acute and/or chronic adverse effects during the kindling acquisition and chronic drug administration period, as well as subsequently during the behavioral testing and drug screening period.

1. Mice are weighed prior to initiating kindling (Day 0), throughout the kindling process prior to LTG/vehicle administration, and then at least weekly throughout the drug screening period when other investigational drugs are administered.

2. In addition to body weight changes, physical health can be assessed using a visual clinical exam scoring system (Table 1; [35, 68, 69]). Each observation session should be performed within the mouse's home cage. Mice should be assessed for their general appearance, behavior, body condition [70], and overall clinical condition by a single investigator blinded to experimental treatment conditions. Within each of those categories, animals are assigned a score of 0–3 based on the visual presence or absence of various abnormal signs or symptoms (Table 1). The scores from each category should be summed to determine an overall clinical score (range of 0–11; [35]).

3. Animals are scored throughout the study at discrete time points to assess behavioral changes during the kindling process and drug administration period until the time of euthanasia. Time points of clinical assessment may include (1) baseline prior to kindling initiation; (2) 1 week into kindling prior to presentation of consistent generalized seizures; (3) 1 week after kindling acquisition, prior to the initial administration of investigational drugs; (4) after 1 week of investigational drug treatment; and (5) after 2–6 weeks of drug treatment. As with kindling itself, care should always be taken to perform behavioral assessments at approximately the same time of day to minimize the potential variability in circadian activity.

3.2 Minimal Motor Impairment Test

To assess the potential for adverse side effects of acute drug administration during drug screening studies, fully kindled mice can be visually evaluated for overt impairments of neurological or muscular function. In mice, the rotarod is often used to identify minimal motor impairment (MMI; [71]). A mouse can maintain its equilibrium for long periods of time on a rod that rotates at a speed of 6 rpm.

Table 1
Clinical assessment parameters and scoring criteria

Score	0	1	2	3
Appearance	Normal	Lack of grooming Sunken eyes Mild blepharitis/ blepharospasm Mild hunched posture	Ocular discharge Corneal ulceration Marked blepharitis/ blepharospasm Prolonged hunched posture Abdominal distention Head tilt	Hunched unmoving posture Labored breathing
Clinical signs	None	Elevated or decreased respiratory and/or effort	Clinical dehydration, marked pallor	
Behavior	Normal	Decreased interaction with cagemates Slow to move about cage Decreased interest in environment	Isolated from cagemates Slow to move when stimulated Impaired mobility Ataxia Hypo- or hyperesthetic when handled	Immobile Weakly or not responsive to handling Convulsing
Body condition	BCS ≥ 3	BCS 2–3	BCS 1–2	BCS ≤ 1

Scoring should be conducted over the course of behavioral testing at approximately the same time every session by an investigator blinded to treatment group. Reprinted with permission [35]

1. The mouse is gently placed on a rotating rod (6 rpm) with all four paws placed parallel to the rotational direction of the rotarod. The animal should be placed so that he can freely and normally "walk" on the rod; e.g., his head is leading the body in the opposite direction that the rod is rotating.

2. The animal is challenged on the rod for 1 min. Any falls are noted during this period. The mouse is considered impaired if it falls off this rotating rod three times during a 1 min period.

3. At the conclusion of the 1 min period or following the third fall (whichever is first), the mouse is returned to the home cage. The animal is then tested for corneal kindled seizure severity with application of bilateral stimulation of the electrical current at the same intensity used throughout the kindling process (e.g., 3 mA, 3 s, 60 Hz stimulation for CF-1 mice).

3.3 Sub-chronic ASM Administration to LTG-Resistant Corneal Kindled Mice

Only LTG-kindled mice with confirmed LTG resistance should be used for drug screening studies and/or behavioral testing.

1. Each investigational drug and dose should be tested in a minimum of $n = 4$ mice/group to determine potential for anticonvulsant efficacy of the candidate compound. Investigational

compounds can be purchased from commercial suppliers and formulated in appropriate diluents, e.g., 0.5% methylcellulose (Sigma, catalogue #M0430). Each investigational compound should be administered by the desired route suited for each candidate compound under study; e.g., compounds administered by the intraperitoneal (i.p.) route are typically administered in a volume of 0.01 mL/g body weight in mice [25].

2. Testing results are recorded and quantified as number of animals (N) protected/the number of mice tested (F). If a median effective (ED50) or median behaviorally impairing dose (TD50) is needed (discussed below), quantification of the ED50/TD50 should be conducted at the time of peak effect of the investigational compound in use [25].

3. Mice should be tested at maximum twice/week with a minimum of 2–3 days' washout between drug administration sessions to minimize the potential for drug-drug interactions (Fig. 1).

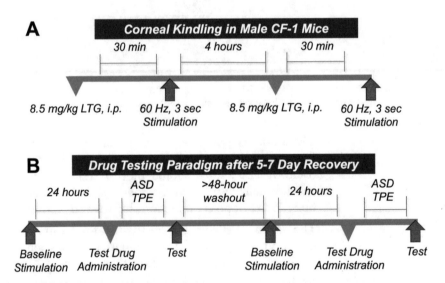

Fig. 1 (**a**) The LTG-resistant corneal kindled mouse is induced by first kindling mice in the presence of an anticonvulsant dose of LTG (8.5 mg/kg, i.p.) throughout the kindling acquisition period. The dose of LTG should be optimized for the mouse strain, sex, and age in use and is based on the median anticonvulsant dose defined in the maximal electroshock test. The current intensity for a 60 Hz 3 s stimulation should be also optimized to the mouse strain, age, and sex in use. Drug administration and kindling should be separated by a minimum of 4 h per daily session during the kindling acquisition process. Mice are kindled to a criterion of five consecutive Racine stage 5 seizures. (**b**) Following a 5–7-day recovery period after achieving the fully kindled state, the LTG sensitivity of mice should be determined in a dose-response challenge (e.g., 2× or 4× the dose of LTG used during kindling acquisition). Fully kindled mice that are determined to be LTG resistant are then candidates for drug administration studies over the course of the following 2–3 months. The actual acute drug evaluation protocol is optimized for moderate-throughput testing within a single week. (Figure adapted with permission [39])

4. The majority of ASM pharmacology in the corneal kindled mouse model has been derived from mice aged 2–5 months old [26, 31, 40, 66]; thus the repeated use of corneal kindled mice is feasible and common practice, but animals should only be retained for repeated drug administration studies up to about 5 months old to minimize the potential for formulation vehicle-induced adverse side effects [35] and drug metabolizing enzyme induction.

3.4 Calculation of the Median Effective (ED50) or Behaviorally Impairing Dose (TD50)

For any rodent seizure and epilepsy model, the ED50 of an investigational compound is the dose needed to suppress a seizure in 50% of the animals in the treatment group [25].

1. The ED50 is determined using the probit method [72] from binary outcome measures ("protected"/"not protected"). Dose selection is typically best achieved by first identifying activity in a range of doses (e.g., logarithmic steps) at the time of peak activity of the specific investigational compound. The same principles can be applied to calculate behaviorally impairing doses (e.g., TD50) or lethal doses (e.g., LD50) for general determination of safety margin of a compound (i.e., protective index, PI, or TD50/ED50; [66]).

2. To calculate an ED50/TD50, group sizes should be appropriately powered; i.e., 8 animals per group give 80% power at 95% significance to detect a difference between the groups that are 1.5-fold greater than the determined standard deviation. In our experience, 5–6 groups ($n = 8$/group) are sufficient to accurately calculate an ED50 or TD50 [66] with 95% confidence intervals within the dose range tested. Wherever possible, pharmacokinetic information (e.g., plasma concentrations of API) should be correlated to pharmacodynamic endpoints (e.g., protection from seizures), as well.

3. The candidate doses of the investigational compound are then administered to groups of $n = 8$ mice, and animals are then tested for severity of the kindled seizure at the time of peak activity of the compound. "Protected" and "not protected" are the primary outcome measures, with seizure scores ≤ 2 considered "protected." The secondary outcome measure is the seizure severity, as measured on the Racine scale.

4. The dose range is modified until at least two points are established between the limits of 0% and 100% protection from seizures. If the probit method is used to calculate ED50, it is essential that the 95% confidence intervals are within the dose range tested. ED50 calculations based on behavioral testing may be conducted over the course of several days.

3.5 Data Acquisition and Interpretation
For kindling acquisition (i.e., during the kindling process itself), the seizure severity on a Racine scale is the primary study outcome measure. The number of stimulations required to achieve the fully kindled state (i.e., five consecutive stage 5 seizures) and the percentage of fully kindled animals within each treatment group are also an important secondary outcome measure.

Following confirmation of the LTG-resistant state, the seizure severity in the presence of any investigational drug is the primary study outcome measure when testing acute ASM efficacy in fully kindled mice. As discussed earlier, kindled seizures are scored according to the Racine scale, with seizure scores ≤ 2 considered "protected." The mean seizure score and standard error of the mean (S.E.M.) of mice in each drug treatment group are calculated.

4   Notes

1. During the kindling process, care should be taken to ensure that the same corneal stimulation electrode location and pressure is applied to every animal to minimize variation in current conductance (i.e., if stimulation is applied via an open or closed eye). Care should also always be taken to perform corneal kindling and subsequent behavioral assessments at approximately the same time of day to minimize variability in responses. Lastly, the same investigator should perform all studies with a given cohort of animals throughout the period of performance to accommodate any potential differences in electrode placement.

2. The number of twice-daily stimulations required to elicit a fully kindled state is entirely dependent on mouse strain, age, and sex. The most important consideration when empirically determining the electrical stimulation necessary for corneal kindling is to ensure that the stimulation intensity is sufficiently low that it does not induce a generalized seizure in the initial phases of kindling (i.e., seizure scores should initially be ≤ 2), but that there is an eventual progression in seizure severity. In our experience, 60 Hz stimulations in the 1.5–3.0 mA range are sufficient to be initially benign and reliably produce a kindled state for most mouse strains across a variety of age ranges and sexes [40].

5   Conclusions

The LTG-resistant corneal kindled mouse provides a relevant platform on which to interrogate mechanism of epilepsy and epileptogenesis, including mechanisms of drug-resistant epilepsy. For example, inflammation may contribute to seizure induction and maintenance [73], and corneal kindled mice are increasingly recognized to exhibit neuroinflammation and neuropathology associated with human epilepsy [32, 33, 74]. However, induction of peripheral inflammation after corneal kindling acquisition through repeated administration of the common formulation vehicle, 0.5% MC, as in the case of the LTG-resistant corneal kindled mouse protocol presently described, does not significantly affect behavior, seizure stability, or presentation up to 6 weeks after corneal kindling [35]. This inflammation induced by repeated, low concentration administration of MC to corneal kindled mice, as is routinely performed during drug evaluation studies with LTG-resistant mice, did not adversely impact the ability of corneal kindled mice to reproduce behavioral facets consistent with clinical epilepsy [35]. These findings are also important to consider due to the fact that systemic administration of the pro-inflammatory agent, lipopolysaccharide, can accelerate kindling acquisition in juvenile rats [75, 76], suggesting that increased inflammation could generally accelerate kindling development or affect the model phenotype. Thus, the investigator should be cautioned that the LTG-resistant corneal kindled mouse model carries the potential to demonstrate histologic lesions as a result of the repeated administration of formulation vehicles, but that this peripheral inflammation due to repeated MC administration does not adversely affect the behavioral aspects of the LTG-resistant corneal kindled mouse model itself.

The pharmacological and behavioral profile of LTG-resistant corneal kindled mouse [39], as well as the ease in which large numbers of mice with uniform seizure history can be rapidly generated in a short time period, demonstrates the suitability of this model for medium-throughput frontline screening of potential ASMs *before* they are advanced to more resource-intensive differentiation models with chronic drug-resistant seizures, e.g., LTG-resistant kindled rat, rats with spontaneous seizures, or focal kainic acid mouse models [10]. Low-dose LTG administration during kindling does not modify acquisition of the fully kindled state, but does lead to a LTG-resistant corneal kindled mouse that is also resistant to other ASMs, including CBZ, VPA, and RTG, consistent with the more labor-intensive LTG-resistant amygdala-kindled rat [56, 77]. As such, the LTG-resistant corneal kindled mouse is a highly useful, moderate-throughput model of pharmacoresistant seizures that is amenable to the early screening of large libraries of investigational compounds for potential applicability to drug-resistant epilepsy. Whether novel therapies that are identified in the LTG-resistant corneal kindled mouse, or similar LTG-resistant kindled rat models, will be found safe and effective in human patients with drug-resistant epilepsy remains to be further defined.

Acknowledgments

This work was supported by the University of Washington Department of Pharmacy and an ITHS KL2 Career Development Award (NCATS 3KL2TR002317).

References

1. Chen Z, Brodie MJ, Liew D et al (2018) Treatment outcomes in patients with newly diagnosed epilepsy treated with established and new antiepileptic drugs: a 30-year longitudinal cohort study. JAMA Neurol 75(3):279–286. https://doi.org/10.1001/jamaneurol.2017.3949

2. Tang F, Hartz AMS, Bauer B (2017) Drug-resistant epilepsy: multiple hypotheses, few answers. Front Neurol 8:301. https://doi.org/10.3389/fneur.2017.00301

3. Loscher W, Klitgaard H, Twyman RE et al (2013) New avenues for anti-epileptic drug discovery and development. Nat Rev Drug Discov 12(10):757–776. https://doi.org/10.1038/nrd4126

4. Modi AC, Ingerski LM, Rausch JR et al (2012) White coat adherence over the first year of therapy in pediatric epilepsy. J Pediatr 161(4):695–699 e691. https://doi.org/10.1016/j.jpeds.2012.03.059

5. Eddy CM, Rickards HE, Cavanna AE (2011) The cognitive impact of antiepileptic drugs. Ther Adv Neurol Disord 4(6):385–407. https://doi.org/10.1177/1756285611417920

6. Loscher W, Potschka H (2002) Role of multidrug transporters in pharmacoresistance to antiepileptic drugs. J Pharmacol Exp Ther 301(1):7–14

7. Begley CE, Durgin TL (2015) The direct cost of epilepsy in the United States: a systematic review of estimates. Epilepsia 56(9):1376–1387. https://doi.org/10.1111/epi.13084

8. Nogueira MH, Yasuda CL, Coan AC et al (2017) Concurrent mood and anxiety disorders are associated with pharmacoresistant seizures in patients with MTLE. Epilepsia 58(7):1268–1276. https://doi.org/10.1111/epi.13781

9. Devinsky O, Spruill T, Thurman D et al (2016) Recognizing and preventing epilepsy-related mortality: a call for action. Neurology 86(8):779–786. https://doi.org/10.1212/WNL.0000000000002253

10. Kehne JH, Klein BD, Raeissi S et al (2017) The National Institute of Neurological Disorders and Stroke (NINDS) Epilepsy Therapy Screening Program (ETSP). Neurochem Res. https://doi.org/10.1007/s11064-017-2275-z

11. Barker-Haliski M, White HS (2019) Validated animal models for antiseizure drug (ASD) discovery: advantages and potential pitfalls in ASD screening. Neuropharmacology:107750. https://doi.org/10.1016/j.neuropharm.2019.107750

12. Putnam TJ, Merritt HH (1937) Experimental determination of the anticonvulsant properties of some phenyl derivatives. Science 85(2213):525–526. https://doi.org/10.1126/science.85.2213.525

13. White HS, Smith-Yockman M, Srivastava A et al (2006) Therapeutic assays for the identification and characterization of antiepileptic and antiepileptogenic drugs. In: Pitkanen A, Schwartzkroin PA, Moshe SL (eds) Models of seizure and epilepsy, 1st edn. Elsevier Academic Press, Burlington, pp 539–549

14. Loscher W (2016) Fit for purpose application of currently existing animal models in the discovery of novel epilepsy therapies. Epilepsy Res 126:157–184. https://doi.org/10.1016/j.eplepsyres.2016.05.016

15. Everett GM, Richards RK (1944) Comparative anticonvulsive action of 3,5,5-trimethyloxazolidine-2,4-dione (Tridione), Dilantin and phenobarbital. J Pharmacol Exp Ther 81:402–407

16. Lennox WG (1945) The petit mal epilepsies. Their treatment with Tridione. JAMA 129:1069–1074

17. Smith M, Wilcox KS, White HS (2007) Discovery of antiepileptic drugs. Neurotherapeutics 4(1):12–17. https://doi.org/10.1016/j.nurt.2006.11.009

18. Barker-Haliski M, White HS (2015) Antiepileptic drug development and experimental models. In: Wyllie E, Gidal BE, Goodkin HP (eds) Wyllie's treatment of epilepsy, 6th edn. Lippencott, Williams & Wilkins, Philadelphia

19. White HS, Loscher W (2014) Searching for the ideal antiepileptogenic agent in experimental models: single treatment versus combinatorial treatment strategies. Neurotherapeutics 11 (2):373–384. https://doi.org/10.1007/s13311-013-0250-1

20. Lothman EW, Hatlelid JM, Zorumski CF et al (1985) Kindling with rapidly recurring hippocampal seizures. Brain Res 360(1–2):83–91

21. McNamara JO, Byrne MC, Dasheiff RM et al (1980) The kindling model of epilepsy: a review. Prog Neurobiol 15:139–159

22. Klitgaard H (2001) Levetiracetam: the preclinical profile of a new class of antiepileptic drugs? Epilepsia 42(Suppl 4):13–18

23. Klitgaard H, Matagne A, Gobert J et al (1998) Evidence for a unique profile of levetiracetam in rodent models of seizures and epilepsy. Eur J Pharmacol 353(2–3):191–206

24. Lothman EW, Williamson JM (1994) Closely spaced recurrent hippocampal seizures elicit two types of heightened epileptogenesis: a rapidly developing, transient kindling and a slowly developing, enduring kindling. Brain Res 649:71–84

25. Barker-Haliski M, Harte-Hargrove LC, Ravizza T et al (2018) A companion to the preclinical common data elements for pharmacologic studies in animal models of seizures and epilepsy. A report of the TASK3 Pharmacology Working Group of the ILAE/AES Joint Translational Task Force. Epilepsia Open 3(Suppl 1):53–68. https://doi.org/10.1002/epi4.12254

26. Rowley NM, White HS (2010) Comparative anticonvulsant efficacy in the corneal kindled mouse model of partial epilepsy: correlation with other seizure and epilepsy models. Epilepsy Res 92(2–3):163–169

27. Grabenstatter HL, Ferraro DJ, Williams PA et al (2005) Use of chronic epilepsy models in antiepileptic drug discovery: the effect of topiramate on spontaneous motor seizures in rats with kainate-induced epilepsy. Epilepsia 46 (1):8–14. https://doi.org/10.1111/j.0013-9580.2005.13404.x

28. Goddard GV, McIntyre DC, Leech CK (1969) A permanent change in brain function resulting from daily electrical stimulation. Exp Neurol 25(3):295–330

29. Goddard GV (1967) Development of epileptic seizures through brain stimulation at low intensity. Nature 214(92):1020–1021

30. Sangdee P, Turkanis SA, Karler R (1982) Kindling-like effect induced by repeated corneal electroshock in mice. Epilepsia 23 (5):471–479

31. Matagne A, Klitgaard H (1998) Validation of corneally kindled mice: a sensitive screening model for partial epilepsy in man. Epilepsy Res 31(1):59–71

32. Loewen JL, Barker-Haliski ML, Dahle EJ et al (2016) Neuronal injury, gliosis, and glial proliferation in two models of temporal lobe epilepsy. J Neuropathol Exp Neurol 75 (4):366–378. https://doi.org/10.1093/jnen/nlw008

33. Remigio GJ, Loewen JL, Heuston S et al (2017) Corneal kindled C57BL/6 mice exhibit saturated dentate gyrus long-term potentiation and associated memory deficits in the absence of overt neuron loss. Neurobiol Dis 105:221–234. https://doi.org/10.1016/j.nbd.2017.06.006

34. Barker-Haliski ML, Vanegas F, Mau MJ et al (2016) Acute cognitive impact of antiseizure drugs in naive rodents and corneal-kindled mice. Epilepsia 57(9):1386–1397. https://doi.org/10.1111/epi.13476

35. Meeker S, Beckman M, Knox KM et al (2019) Repeated intraperitoneal administration of low-concentration methylcellulose leads to systemic histologic lesions without loss of preclinical phenotype. J Pharmacol Exp Ther. https://doi.org/10.1124/jpet.119.257261

36. Rogawski MA (2006) Diverse mechanisms of antiepileptic drugs in the development pipeline. Epilepsy Res 69(3):273–294. https://doi.org/10.1016/j.eplepsyres.2006.02.004

37. Klein P, Dingledine R, Aronica E et al (2018) Commonalities in epileptogenic processes from different acute brain insults: do they translate? Epilepsia 59(1):37–66. https://doi.org/10.1111/epi.13965

38. Barker-Haliski M (2019) How do we choose the appropriate animal model for antiseizure therapy development? Expert Opin Drug Discovery:1–5. https://doi.org/10.1080/17460441.2019.1636782

39. Koneval Z, Knox KM, White HS et al (2018) Lamotrigine-resistant corneal-kindled mice: a model of pharmacoresistant partial epilepsy for moderate-throughput drug discovery. Epilepsia 59(6):1245–1256. https://doi.org/10.1111/epi.14190

40. Beckman M, Knox K, Koneval Z et al (2019) Loss of presenilin 2 age-dependently alters susceptibility to acute seizures and kindling acquisition. Neurobiol Dis 136:104719. https://doi.org/10.1016/j.nbd.2019.104719

41. Engstrom FL, White HS, Kemp JW et al (1986) Acute and chronic acetazolamide administration in DBA and C57 mice: effects of age. Epilepsia 27(1):19–26

42. Frankel WN, Taylor L, Beyer B et al (2001) Electroconvulsive thresholds of inbred mouse strains. Genomics 74(3):306–312

43. Leclercq K, Kaminski RM (2015) Genetic background of mice strongly influences treatment resistance in the 6 Hz seizure model. Epilepsia 56(2):310–318. https://doi.org/10.1111/epi.12893

44. Koneval Z, Knox KM, Memon A et al. (2020) Antiseizure drug efficacy and tolerability in established and novel drug discovery seizure models in outbred versus inbred mice. Epilepsia Sep;61(9):2022–2034

45. Otto JF, Singh NA, Dahle EJ et al (2009) Electroconvulsive seizure thresholds and kindling acquisition rates are altered in mouse models of human KCNQ2 and KCNQ3 mutations for benign familial neonatal convulsions. Epilepsia 50(7):1752–1759. https://doi.org/10.1111/j.1528-1167.2009.02100.x

46. Otto JF, Yang Y, Frankel WN et al (2004) Mice carrying the szt1 mutation exhibit increased seizure susceptibility and altered sensitivity to compounds acting at the m-channel. Epilepsia 45(9):1009–1016. https://doi.org/10.1111/j.0013-9580.2004.65703.x

47. Goodman LS, Grewal MS, Brown WC et al (1953) Comparison of maximal seizures evoked by pentylenetetrazol (metrazol) and electroshock in mice, and their modification by anticonvulsants. J Pharmacol Exp Ther 108(2):168–176

48. Kilkenny C, Browne W, Cuthill IC et al (2011) Animal research: reporting in vivo experiments – the ARRIVE guidelines. J Cereb Blood Flow Metab 31(4):991–993. https://doi.org/10.1038/jcbfm.2010.220

49. Harte-Hargrove LC, French JA, Pitkanen A et al (2017) Common data elements for preclinical epilepsy research: standards for data collection and reporting. A TASK3 report of the AES/ILAE Translational Task Force of the ILAE. Epilepsia 58(Suppl 4):78–86. https://doi.org/10.1111/epi.13906

50. Loscher W, Fassbender CP, Nolting B (1991) The role of technical, biological and pharmacological factors in the laboratory evaluation of anticonvulsant drugs. II Maximal electroshock seizure models. Epilepsy Res 8(2):79–94

51. Leclercq K, Matagne A, Kaminski RM (2014) Low potency and limited efficacy of antiepileptic drugs in the mouse 6 Hz corneal kindling model. Epilepsy Res 108(4):675–683. https://doi.org/10.1016/j.eplepsyres.2014.02.013

52. Woodbury L, Davenport V (1952) Design and use of a new electroshock seizure apparatus, and analysis of factors altering seizure threshold and pattern. Arch Int Pharmacodyn Ther 92:97–104

53. Bialer M, Twyman RE, White HS (2004) Correlation analysis between anticonvulsant ED50 values of antiepileptic drugs in mice and rats and their therapeutic doses and plasma levels. Epilepsy Behav 5(6):866–872

54. Cela E, McFarlan AR, Chung AJ et al (2019) An optogenetic kindling model of neocortical epilepsy. Sci Rep 9(1):5236. https://doi.org/10.1038/s41598-019-41533-2

55. Krug M, Koch M, Grecksch G et al (1997) Pentylenetetrazol kindling changes the ability to induce potentiation phenomena in the hippocampal CA1 region. Physiol Behav 62(4):721–727

56. Srivastava AK, White HS (2013) Carbamazepine, but not valproate, displays pharmacoresistance in lamotrigine-resistant amygdala kindled rats. Epilepsy Res 104(1–2):26–34. https://doi.org/10.1016/j.eplepsyres.2012.10.003

57. Srivastava AK, Woodhead JH, White HS (2003) Effect of lamotrigine, carbamazepine, and sodium valproate on lamotrigine-resistant kindled rats. Epilepsia 44:42

58. Singh E, Pillai KK, Mehndiratta M (2014) Characterization of a lamotrigine-resistant kindled model of epilepsy in mice: evaluation of drug resistance mechanisms. Basic Clin Pharmacol Toxicol 115(5):373–378. https://doi.org/10.1111/bcpt.12238

59. Sato M, Racine RJ, McIntyre DC (1990) Kindling: basic mechanisms and clinical validity. Electroencephalogr Clin Neurophysiol 76(5):459–472

60. Loscher W, Rundfeldt C, Honack D (1993) Pharmacological characterization of phenytoin-resistant amygdala- kindled rats, a new model of drug-resistant partial epilepsy. Epilepsy Res 15(3):207–219

61. Loscher W, Brandt C (2010) Prevention or modification of epileptogenesis after brain insults: experimental approaches and translational research. Pharmacol Rev 62(4):668–700. https://doi.org/10.1124/pr.110.003046

62. Loscher W, Honack D, Rundfeldt C (1998) Antiepileptogenic effects of the novel anticonvulsant levetiracetam (ucb L059) in the kindling model of temporal lobe epilepsy. J Pharmacol Exp Ther 284(2):474–479

63. Otsuki K, Morimoto K, Sato K et al (1998) Effects of lamotrigine and conventional antiepileptic drugs on amygdala- and hippocampal-kindled seizures in rats. Epilepsy Res 31(2):101–112

64. Wauquier A, Zhou S (1996) Topiramate: a potent anticonvulsant in the amygdala-kindled rat. Epilepsy Res 24(2):73–77

65. Loscher W, Honack D (1993) Profile of ucb L059, a novel anticonvulsant drug, in models of partial and generalized epilepsy in mice and rats. Eur J Pharmacol 232(2–3):147–158

66. Barker-Haliski ML, Johnson K, Billingsley P et al (2017) Validation of a preclinical drug screening platform for pharmacoresistant epilepsy. Neurochem Res 42(7):1904–1918. https://doi.org/10.1007/s11064-017-2227-7

67. Racine RJ (1972) Modification of seizure activity by electrical stimulation: II. Motor seizure. Electroencephalogr Clin Neurophysiol 32:281–294

68. Rex TS, Boyd K, Apple T et al (2016) Effects of repeated anesthesia containing urethane on tumor formation and health scores in male C57BL/6J mice. J Am Assoc Lab Anim Sci 55(3):295–299

69. Mai SHC, Sharma N, Kwong AC et al (2018) Body temperature and mouse scoring systems as surrogate markers of death in cecal ligation and puncture sepsis. Intensive Care Med Exp 6 (1):20. https://doi.org/10.1186/s40635-018-0184-3

70. Ullman-Cullere MH, Foltz CJ (1999) Body condition scoring: a rapid and accurate method for assessing health status in mice. Lab Anim Sci 49(3):319–323

71. Dunham MS, Miya TA (1957) A note on a simple apparatus for detecting neurological deficit in rats and mice. J Am Pharm Assoc Sci Ed 46:208–209

72. Finney DJ (1952) Probit analysis. A statistical treatment of the sigmoid response curve. University Press, Cambridge

73. Vezzani A, Friedman A, Dingledine RJ (2013) The role of inflammation in epileptogenesis. Neuropharmacology 69:16–24. https://doi.org/10.1016/j.neuropharm.2012.04.004

74. Albertini G, Walrave L, Demuyser T et al (2017) 6 Hz corneal kindling in mice triggers neurobehavioral comorbidities accompanied by relevant changes in c-Fos immunoreactivity throughout the brain. Epilepsia. https://doi.org/10.1111/epi.13943

75. Auvin S, Shin D, Mazarati A et al (2010) Inflammation induced by LPS enhances epileptogenesis in immature rat and may be partially reversed by IL1RA. Epilepsia 51(Suppl 3):34–38. https://doi.org/10.1111/j.1528-1167.2010.02606.x

76. Auvin S, Mazarati A, Shin D et al (2010) Inflammation enhances epileptogenesis in the developing rat brain. Neurobiol Dis 40 (1):303–310. https://doi.org/10.1016/j.nbd.2010.06.004

77. Metcalf CS, Huff J, Thomson KE et al (2019) Evaluation of antiseizure drug efficacy and tolerability in the rat lamotrigine-resistant amygdala kindling model. Epilepsia Open 4 (3):452–463. https://doi.org/10.1002/epi4.12354

Chapter 11

Protocol for Drug Screening with Quantitative Video-Electroencephalography in a Translational Model of Refractory Neonatal Seizures

Brennan J. Sullivan and Shilpa D. Kadam

Abstract

The neonatal brain is more prone to seizures than the adult brain, with the highest incidence of seizures occurring during the first year of life. Neonatal seizures are the most common neurological emergency in newborns, occurring in 1–3.5 per 1000 newborns. The majority of neonatal seizures, chiefly those associated with hypoxic ischemic encephalopathy (HIE), are refractory to first-line anti-seizure medications (ASMs). The global first-line ASM for HIE neonatal seizures is phenobarbital (PB), a positive allosteric modulator of $GABA_A$ receptors ($GABA_A Rs$). However, commonly used loading doses of PB are often ineffective for curbing HIE neonatal seizures. Without efficacious management strategies, poor seizure management devolves into prolonged and recurrent seizures associated with significant mortality, neurodevelopmental comorbidities, and epileptogenesis. The mechanisms underlying refractory neonatal seizures are poorly understood, as preclinical models that recapitulate important aspects of HIE historically have not been validated for their acute seizure burdens or ASM response quantitatively with video-electroencephalography (vEEG). In this chapter, we describe a protocol for drug screening with quantitative vEEG in a translationally sound model of refractory neonatal seizures.

Key words Quantitative vEEG, Refractory neonatal seizures, Translational seizure model

Abbreviations

ASM	Anti-seizure medication
BREDs	Brief runs of epileptiform discharges
$GABA_A R$	$GABA_A$ receptor
$GABA_B R$	$GABA_B$ receptor
HIE	Hypoxic ischemic encephalopathy
KCC2	K^+-Cl^- cotransporter type 2
NICU	Neonatal intensive care unit
P	Postnatal day
PB	Phenobarbital

Divya Vohora (ed.), *Experimental and Translational Methods to Screen Drugs Effective Against Seizures and Epilepsy*, Neuromethods, vol. 167, https://doi.org/10.1007/978-1-0716-1254-5_11,

PTZ   Pentylenetetrazol
TrkB   Tyrosine receptor kinase B
vEEG   Video-electroencephalograph

---

1 Introduction

Efficacious management of acute neonatal seizures is notoriously elusive at the bedside. Neonatal seizures are difficult to diagnose; clinical manifestations range from non-convulsive to movements that can be easily misinterpreted as benign [1]. The majority of HIE neonatal seizures are sub-clinical and do not initially present with discernible convulsive manifestations, requiring obligatory EEG monitoring for diagnosis [2]. The timely diagnosis of neonatal seizures after HIE is variable, partially due to their self-limiting nature [3]. Further, neonatal seizures that initially present with clinical manifestations often become sub-clinical after ASM intervention [4]. Thus, continuous vEEG is the gold standard for diagnosing neonatal seizures, evaluating seizure severity, and determining ASM efficacy [1, 2].

There is no consensus on a standardized protocol for treating refractory neonatal seizures in HIE after the first loading dose of PB fails. The specific criteria for determining a neonate to be high-risk, availability of subsequent continuous EEG monitoring, diagnosis of neonatal seizures, timely ASM intervention, and optimal EEG recording necessary to confirm ASM efficacy differ greatly between neonatal intensive care units (NICUs) [5]. Additionally, neonatal seizures have varying temporal onsets after HIE with seizure burdens that wax and wane over the first week of life. Therefore, the administration of ASMs during the trough or peak periods of seizure burden could elicit disparate results and false positives [6]. As a consequence of these known complexities, there is an urgent need for clinical trial designs that accommodate for the phenotype of neonatal seizures [7]. In contrast, preclinical experiments at the bench are able to systematically investigate the time course, ASM efficacy, and mechanisms underlying neonatal seizures. Without insights gained from preclinical research, the urgent and unmet need for evidence-based management for refractory neonatal seizures is unlikely to improve [8].

1.1 Design Rationale

The value of any model for seizures or epileptogenesis is its ability to investigate the mechanisms underlying pathogenesis for the overall purpose of developing treatments that have potential to translate into effective therapies for patients [9]. Designing and investigating an efficient therapy for treating refractory neonatal seizures is challenging because there are intrinsic difficulties in dissociating the harmful effects between hypoxic ischemia and recurrent refractory seizures. In this regard, the indispensable qualification for a model of HIE neonatal seizures is refractoriness to first-line ASMs.

Preclinical models have been developed to examine neonatal seizures within in vivo and in vitro preparations that utilize chemoconvulsants, prolonged hypoxic conditions, and/or ischemia (for review *see* Kang and Kadam [10]). The CD-1 ischemia model of HIE neonatal seizures holds transitional potential as it provides robust seizure burdens, refractoriness to a loading dose of PB, and transient seizures during development with long-term neurological comorbidities [11, 12]. Some neonatal seizure models may not meet this translational standard due to low seizure burdens and the failure to demonstrate ASM refractoriness. For example, neonatal seizures induced by chemoconvulsants respond to ASMs and therefore lack the unique pathogenesis of refractory seizures following ischemia [13]. Clinical trials for novel interventions in HIE were initiated on the basis of results derived from chemoconvulsant neonatal seizure models; the initial clinical trial was terminated early and found no efficacy [14]. Preclinical investigations of these same novel interventions were tested in the ischemia model of neonatal refractory seizures and also demonstrated a lack of efficacy [11].

Models of pediatric acute seizures or epilepsy must recognize the developmental neurobiology underlying the age of seizure susceptibility. Rodents are currently the most widely used model organism for studies of in vivo seizures and chronic epilepsy. In spite of species differences with humans, rodent models have provided novel insights into the human pathophysiology of seizures and epilepsy [9]. However, directly comparing the developmental stages between rodents and humans is a non-trivial task that requires the characterization of neurobiological developmental stages and processes across species. For example, in the mature healthy mammalian central nervous system, the neurotransmitter GABA mediates a net inhibitory effect on neurons via $GABA_ARs$ or GABAB receptors (GABABRs). The efficacy of $GABA_AR$-mediated fast synaptic inhibition is heavily influenced by the ability of neurons to maintain a low intracellular Cl^- concentration (for review *see* Doyon et al. [15]). During development, low intracellular Cl^- concentrations emerge with an increase in the neuron-specific chloride extruder K^+-Cl^- cotransporter type 2 (KCC2; for review *see* Kaila et al. [16]). In our CD-1 mouse model of HIE, pups are refractory to PB following unilateral carotid ligation at P7 but not at P10, when KCC2 expression is ~200% greater than at P7 [11]. In contrast to mice and rats, the guinea pig has high levels of KCC2 before birth [17]; thus investigation of species or strain-specific developmental trajectories is essential for neonatal seizure models.

When comparing the developmental stages between rodents and humans, lessons can be learned from the epilepsy-associated genome. Advances in the understanding of genetic epilepsies have identified multiple causative de novo dominant mutations that result in haploinsufficiency, refractory seizures, and epilepsy [18]. Heterozygous deletions of proposed epilepsy genes in mice have illustrated complex differences in gene dosing within animal models [19]. This is strikingly apparent in preclinical studies that investigate monogenetic causes of pediatric epilepsy. Many preclinical mouse models of genetic epilepsy do not exhibit spontaneous seizures in mature mice [20, 21], whereas human patients with similar mutations can be diagnosed with early-onset epileptic encephalopathy [22]. One course of action for preclinical research investigating proposed monogenetic causes of pediatric epilepsy is to identify spontaneous seizures and establish the natural history of epilepsy in adult mice [23]. Establishing the adult phenotype provides the investigator with a phenotypic checkpoint to evaluate the developmental trajectories and epilepsy at neonatal and juvenile ages. Taken together, thoughtful design is critical when investigating a novel therapy for treating refractory seizures.

2 The Mechanics

Seizure detection in neonatal mouse models requires an understanding of pup behavior during different stages of development. Well-designed experiments that reflect an understanding of neonatal behavior and emulate home-cage conditions allow high-quality EEG recordings. Some of the critical points to consider are:

1. Neonatal mouse pups weigh ~1–8 g depending on the age, strain, and sex. Therefore, any neonatal EEG recording paradigm cannot impose significant weight on the head or scalp of pups.

2. Mice are born with eyes closed, and all the exploratory movements in the neonatal age (first 2 weeks of life) are related to feeding and seeking warmth, as pups do not effectively autoregulate body temperature. Therefore, litters of neonatal pups are huddled together within a nest provided by the dam. The EEG recording chamber needs to be temperature regulated to avoid excessive stress, hypothermia, and excessive movement artifacts on EEG.

3. At P7 the neonatal skull is non-ossified, thin, and fragile. The parietal bone plates are flexible and cannot maintain their convexity at the sagittal suture when challenged by pressure or external manipulation. These conditions occur when attempting to implant electrodes for epidural, subdural, or depth electrode recording.

4. Manipulations of the unstable neonatal skull during implantation of electrodes can lead to transient increases in intracranial pressure leading to periventricular bleeds.

5. At P7 the periosteum is fragile; therefore trying to tether electrodes to the perichondrium/periosteum is not recommended (having tried and failed ourselves).

6. Seizures involve motor movements that can result in an entanglement of both implanted electrodes and the tether; thus a rotating commutator and adequate tether length are essential (Fig. 1).

7. In contrast to humans, the scalp of neonatal pups is loosely attached to the skull, allowing for easy implantation of subdermal scalp electrodes.

8. Subdermal scalp electrodes have resulted in robust, reproducible, and artifact-limited neonatal mouse pup recordings (Fig. 2).

All of the points mentioned above should be considered prior to recording the EEG, as there are no post-hoc methods that produce reliable signals from EEG recordings with large sources of artifacts and low signal-to-noise ratios.

3   Methods and Data Acquisition

3.1 Animal Acquisition and Housing

All experimental procedures described are conducted in compliance with protocols approved by Johns Hopkins University Animal Care and Use Committee. CD-1 mice are initially purchased from Charles River Laboratories, Inc. (Wilmington, MA). Newly born litters of pups are delivered at postnatal day (P) 3 or P6 and acclimate to P7 or P10, respectively. All litters of mice are single-housed with the dam, providing food and water ad libitum on a 12 h light-dark cycle. It is important to emphasize that acute insults from neonatal unilateral carotid ligation in mice are strain and age dependent [11, 24].

3.2 Unilateral Carotid Ligation Surgical Procedure and vEEG Recording

On the day of surgery, each pup is assigned a numerical identifier, sex is determined, and each is subsequently weighed (for step-by-step protocol, see Table 1). Isothermal heating pads are warmed and placed within a recording chamber before surgery to allow the recording chamber to reach a stable temperature of 36 °C. We employ an 8–12 min range as our standard for surgery duration; this ensures a consistent recovery time from anesthesia for each individual pup. Animals are anesthetized with 3.0–1.5% of isoflurane carried by a 50/50 mixture of O_2 and N_2 for the duration of the surgery (8–12 min). During surgery pups are loosely (to avoid restriction of airflow to the lungs) taped down, with the neck

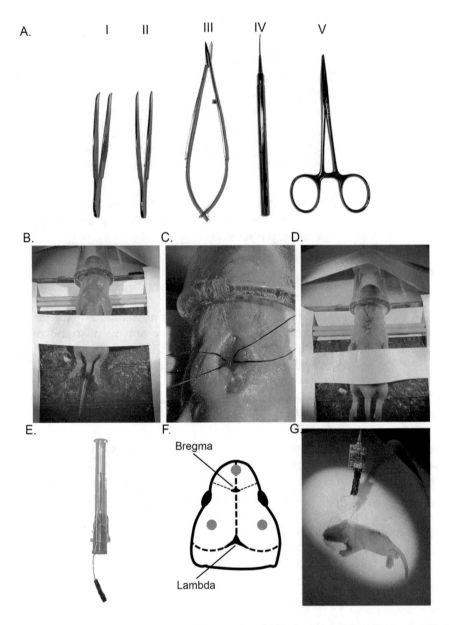

Fig. 1 P7 unilateral carotid ligation and electrode implantation in CD-1 mice. (**a**) Tools for surgery. From left to right: fine-tip blunt forceps (I–II), spring micro-scissors (III), blunt hook (IV), and needle holder (V). (**b**) P7 CD-1 pup under maintenance 1.5% isoflurane on a platform extending the neck. (**c**) Silk placement under right carotid artery. (**d**) Suture. (**e, f**) 5 kΩ electrode and electrode placement (gray dots represent location of electrodes). (**g**) Preamplifier with electrode attachment

extended (Fig. 1). Animals are then subjected to permanent unilateral ligation of the right common carotid artery using 6-0 surgisilk (Fine Science Tools, BC, Canada), and the outer skin is subsequently closed with 6-0 monofilament nylon (Covidien, MA, USA). The local anesthetic lidocaine is applied after suturing (Fig. 1).

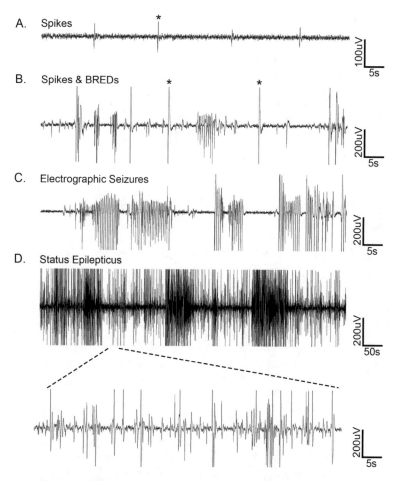

Fig. 2 Representative EEG traces in neonatal pups after unilateral carotid ligation. EEG traces from a CD-1 pups post-ligation demonstrate spiking (**a**, **b**), electrographic brief runs of epileptiform discharges (BREDs) (**c**), and status epilepticus (**d**). EEG of status epilepticus, with expansion EEG showing continuous activity that does not return to baseline (see inset; * denotes interictal spikes)

Pups are then implanted with three subdermal EEG electrodes (one recording, one reference, and one ground) on the skull overlying the ipsilateral and contralateral parietal cortex and rostrum (Fig. 1). Subsequently, pups are tethered and placed within the isothermal vEEG recording chamber during recovery from anesthesia. Sham control animals are treated identically (i.e., 8 min of isoflurane anesthesia, incision, suture, etc.) except for the carotid ligation, to control for strain and anesthesia-specific EEG modulations (for review *see* Kadam et al. [25]). Once pups recover from anesthesia (2–5 min), vEEG recording begins (Fig. 1). At the end of the recording session, subdermal electrodes are removed from the pups, and then the pups are immediately returned to the damn

Table 1
Neonatal unilateral carotid ligation protocol: step-by-step methodology

Preparation
1. Make drug solutions for injection on day of experiment (i.e., phenobarbital)
2. Collect autoclaved surgical instruments
3. Prepare electrodes
4. Prepare Hamilton syringes for loading drug doses
5. Wear surgical gloves, gown, and face mask
6. Cut surgical silk into ~3″ segments (one for each pup)
7. Warm isothermal heating pads, and allow each recording chamber to reach 36 °C
8. Record pup sex and weight for the litter
Surgery
1. Place pup supine on surgery platform and insert nose into the anesthesia cone
2. Initiate induction anesthesia (3% isoflurane)
3. Secure pup with tape with slight neck extension
4. Apply surgical scrub (Betadine) to neck and let dry
5. Before initiating skin incision, check toe-pinch reflex for depth of anesthesia
6. At the midline of the neck, lift skin with blunted forceps and make a vertical incision with scissors
7. Use blunt forceps to dissect through fat, neck muscles, and fascia to expose the right carotid artery
8. The right carotid artery is identified by anatomical location, cardiac pulsations, and red color
9. Place surgical silk loop lateral to the right carotid artery
10. Using the surgical hook from the opposite side, pull the loop of the silk from under the carotid
11. Cut silk loop with scissors to make two strands
12. Use the lower silk strand to ligate the carotid closer to the heart; then use the upper strand
13. Close and suture incision site with skin good approximation using nylon
14. Apply local anesthetic and turn pup over to prone position 15. Implant subdermal electrodes using access path created by tip of a sterile gauge 21 needle
16. Insert and fix electrodes in position (flat against skull) with cyanoacrylate adhesive 17. Connect three electrode pins into the pin connector 18. Stop anesthesia
Recording vEEG
1. Place pup in recording chamber and connect the pin connector to the preamplifier
2. Start recording EEG after pup has recovered from anesthesia
3. Administer a loading dose of phenobarbital at 1 h of EEG recording
4. End EEG recording at 2 h
5. Remove subdermal electrodes by cutting wire as close to the skin as possible 6. Apply Betadine generously to scalp and let dry and return pup to dam and littermates

after an antiseptic (Betadine) is applied. Tethered recordings with pups isolated from the dam should not exceed 3–4 h as separation from the dam during the neonatal period is known to induce significant stress that influences gene expression and has long-term behavioral impairments [26].

3.3 In Vivo vEEG Acquisition

EEG recordings are acquired using Sirenia Acquisition software with synchronous video capture (Pinnacle Technology, Kansas, USA). Utilizing a three-channel tethered EEG system, referential EEG recordings are acquired from 0.5 to 50 Hz with a 400 sampling rate and 100 gain. Regardless of the EEG acquisition system, a reliable low-impedance electrical connection must be established and maintained during the entire recording. EEG signals are initially processed by preamplifiers, low-gain amplifiers that improve signal transmission and reduce noise. Shortening the distance between the electrodes and preamplifier reduces noise. The rate at which waveform data from EEG are sampled and converted into numerical values is the sampling rate, and it must be at least twice the highest frequency of interest (Nyquist rate [27]). For further guidelines on EEG data acquisition, *see* Moyer et al. [28].

3.4 Off-Line EEG Analysis

After acquisition, EEG traces are scored manually for electrographic parameters and behavioral scores. All scorers are blinded and files are de-identified. Movement artifact is removed by filtering out <3 Hz and analyzing 3–50 Hz. A single neonatal seizure is defined and scored as an evolving EEG pattern that lasts greater than 6 s, with video confirmation of standardized behavioral manifestations.

The behavioral manifestations of the acute post-ischemic seizures in our neonatal CD-1 mouse model are graded on a scale from 0 to 6 (0, motionless/inactive; 1, flexor spasms; 2, jittery movements; 3, repetitive grooming/scratching, circling, or head bobbing; 4, limb clonus, unstable posture; 5, mice that exhibited level 4 behaviors for >30 s; and 6, severe tonic-clonic behavior with inability to regain loss of posture). Seizure grades 0–3 are defined as non-convulsive, while seizure grades 4–6 are defined as convulsive [11]. Detection and quantitative analysis of seizures with neonatal tethered EEG remains dependent upon visuo-manual review and scoring as neonatal seizures vary in their amplitude, duration, and frequency (Fig. 2). The required visuo-manual review resembles current limitations in the NICU. Neonatologists have adopted the use of amplitude-integrated EEGs, a type of EEG with a compressed output that was initially developed for monitoring adults during anesthesia [29]. The amplitude-integrated EEG was accepted as a compromise in the NICU because less resources are needed to initiate and interpret amplitude-integrated EEG than continuous EEG with a 10–20 montage, even though continuous EEG provides superior seizure detection [30, 31]. Ongoing research is addressing the development and implementation of clinically reliable neonatal seizure algorithms from continuous EEG recordings [32].

4 Notes

4.1 Recommendations for Protocol: Data Acquisition and Troubleshooting

1. Keep all handling, recording chamber preparation, time of day for recording, and anesthesia durations identical for experimental and control pups.

2. Maintain a detailed surgery log for notes: the doses, route, and timing of all injections during EEG recording.

3. Ideally every P7 pup litter should have sex-matched littermates as controls (vehicle-treated) with EEG recordings for durations similar to those of the experimental groups.

4. Keep drug solution volumes for IP injection in neonatal pups as small as possible (5–10 µL) to prevent backtracking of drug solution from abdominal pressure.

5. Inject IP slowly while holding pup in a head-lowered angle to reduce abdominal pressure in iliac fossa, and keep pup in this position for 10 s after slow withdrawal of needle.

6. Set sampling frequency for EEG according to the highest frequency of interest using the Nyquist theorem.

7. Synchronized video in this protocol is critical to facilitate identification of movement artifacts that might resemble pathological patterns in EEG recordings.

8. Keep all acquisition and analyses software updated.

9. Back up all data locally and to a secure server.

5 Drug Screening

In comparison to other models and workflows that are designed to accomplish high-throughput screens of potential ASMs [33], our model is designed specifically to address the underlying mechanisms and interventions for refractory neonatal seizures. Our current workflow (See Table 1 and Notes) is labor-intensive but directly examines the efficacy of novel compounds in an etiologically relevant paradigm (Fig. 3). Recording continuous EEG for 2 h allows for each individual pup to act as its own control, comparing the first hour seizure burden to the second hour seizure burden. At P7, 3 h EEG recordings demonstrate that untreated or PB-treated pups after ligation both have stable seizure burdens, highlighting the refractoriness at P7 with consistent seizure burden over 3 h (Fig. 3). This design incorporates natural variability in EEG seizure burdens

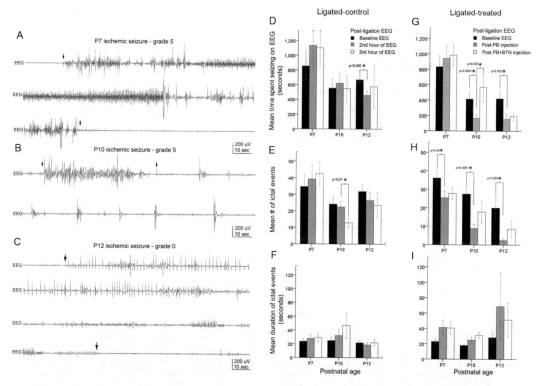

Fig. 3 Age-dependent neonatal refractory seizures in the CD-1 mouse model. (**a–c**) Representative EEG traces of age-dependent seizure phenotypes after unilateral carotid ligation in P7–P12 pups. (**d**) Age-dependent seizure burdens remained stable during 3 h of continuous EEG recording in untreated pups after ligation. (**e, f**) Ictal events and average ictal event duration in untreated pups after ligation. (**g**) P7 pups are refractory to a loading dose of phenobarbital (PB) at 1 h and a subsequent loading dose of PB with bumetanide at 2 h after ligation. P10 and P12 pups have lower initial seizure burdens and are not refractory to PB. (**h, i**) Regardless of age all ictal events decrease in the second hour after the initial loading dose of PB, and this decrease is compensated for by an increase in the average duration of ictal events. (From Kang et al. [11])

between pups and measures ASM efficacy using the same seizing pup as its own control. Individual pups can be graded by initial seizure severity and subsequently stratified into mild, moderate, and severe seizure burdens to evaluate novel compounds with respect to sex, age, and/or seizure severity. To this end, the model has been successful in identifying the ability of the small-molecule tyrosine receptor kinase B (TrkB) antagonist ANA-12 to rescue PB-refractory neonatal seizures [34, 35] (Fig. 4). This rescue is associated with improved long-term outcomes compared to neonatal pups with refractory seizures [12, 36]. The experimental design used to identify ANA-12's dose-dependent efficacy was administration of ANA-12 immediately after ligation with subsequent vEEG for 2 h, with a loading dose of PB (25 mg/kg) at the end of the first hour. This protocol enables the investigation of compounds of interest quantitatively during both the first hour before PB administration and during the second hour after a

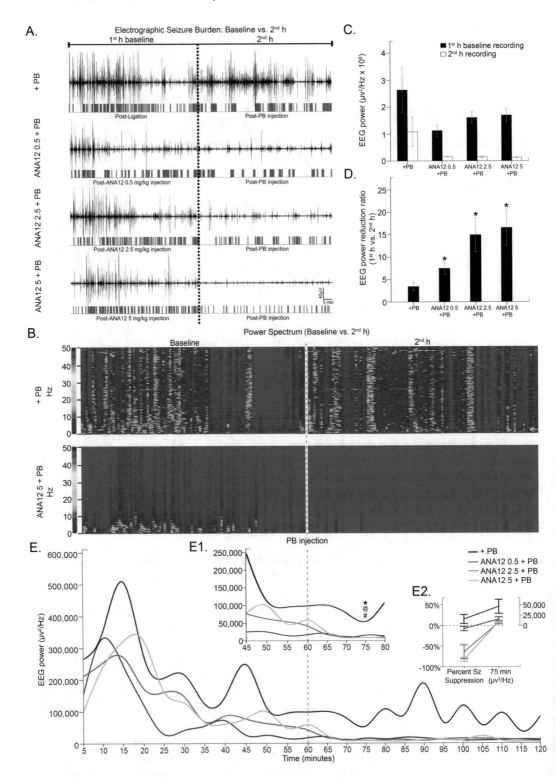

Fig. 4 ANA-12 rescues refractory neonatal seizures in the CD-1 mouse model. (**a**) Representative EEG traces of the dose-dependent rescue of phenobarbital (PB)-refractory neonatal seizures by ANA-12 after unilateral

loading dose of PB. This has enabled the successful identification of ANA-12 and a KCC2 functional enhancer CLP290 to rescue PB-refractory seizures (US Patent #: US20170281579).

6 Etiology of Acute Seizures Dictates Response to Anti-seizure Drugs

Model-specific protocols used to initiate acute seizures can modulate chloride cotransporter expression and function differentially, as well as ASM efficacy [13]. CD-1 P7 neonatal ischemia significantly activates the TrkB pathway in an age- and sex-dependent manner, resulting in TrkB-mediated KCC2 degradation that underlies the emergence of PB refractoriness [11, 37]. In contrast, acute administration of pentylenetetrazol (PTZ) to P7 CD-1 pups induces seizures that are PB-responsive and upregulate KCC213 (Fig. 5). Similar findings of upregulated KCC2 after chemoconvulsant-induced seizures are reported in other models [38–40]. The etiology of neonatal seizures in preclinical models can result in differential effects on KCC2 expression and function; this may underlie the opposite anti-seizure effects of $GABA_AR$ positive allosteric modulators like PB. Historically, chemoconvulsants were used to induce status epilepticus, with the intention of generating spontaneously seizing animals [41, 42]. More recently, chemoconvulsants are being used to induce acute seizures in neonatal models because of their convenience. However, when chemoconvulsants are delivered systemically to assess acute seizures, their extensive cellular and physiological actions (most of which are independent of seizure phenomena) make it challenging to discern the direct effects of the seizures themselves [41]. Therefore, the use of chemoconvulsants has been recommended only for acute focally evoked seizure models [41].

The ability for unilateral carotid ligation alone to induce PB-refractory ischemic seizures is strain specific to CD-1 mice. Studies in other strains of mice and in Sprague Dawley rats require ischemia followed by global hypoxia (the Rice-Vannucci model [43]) to induce seizures [44, 45]. This highlights the strain and species differences in ischemia-induced seizure susceptibility. Additionally, the Rice-Vannucci model for HIE results in a large middle-cerebral-artery-territory necrotic infarcts, which may reflect neonatal stroke and not a hypoxic ischemic encephalopathy [46]. In our P7 CD-1 model of unilateral carotid ligation alone, the histology at P18 shows no necrotic infarcts but diffuse unilateral edema that is transient [11]. These features, associated with the transient

Fig. 4 (continued) carotid ligation in P7 pups (seizure frequency is denoted by the pink raster plot). (**b–e2**) EEG power is not a reliable proxy for EEG seizure scoring, as it is unable to identify the dose-dependent differences in seizure frequency after graded doses of ANA-12. (From Carter et al. [34])

214   Brennan J. Sullivan and Shilpa D. Kadam

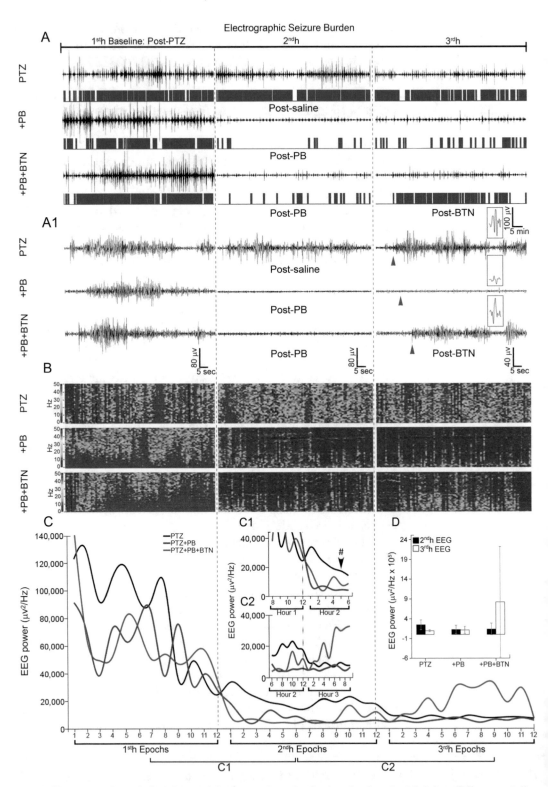

Fig. 5 Chemoconvulsant-induced neonatal seizures are not refractory to phenobarbital. (**a–a1**) Representative EEG traces demonstrate that pentylenetetrazole (PTZ)-induced seizures are not refractory to phenobarbital (PB) at P7 in CD-1 pups. (**b–d**) EEG power is not a reliable proxy for determining seizure burden. (From Kharod et al. [13])

ischemic seizures that disappear with advancing neonatal age, make the model unique to represent HIE pathophysiology both for seizure semiology and differentiate it from neonatal stroke models.

In summary, the method of inducing seizures in preclinical models can result in differential effects on KCC2 expression and function that underlie the efficacy of GABA$_A$R positive allosteric modulators like PB. When the clinical condition being modeled is refractory HIE seizures, the preclinical model should reliably demonstrate refractoriness. Mechanistic etiologies underlying acquired early-life seizures are critical to replicate for translationally viable preclinical modeling.

7   Conclusion

In this chapter, the methods to record vEEGs using a recently characterized model of neonatal ischemic seizures in CD-1 mice are described. Utilizing this model, the seizure susceptibility to a unilateral ischemic insult at P7, P10, and P12 has been established. The emergence of refractory seizures at P7, with PB-responsive seizures at P10 and P12, indicates that the switch in PB efficacy occurs between the ages of P7 and P10. Comparing and contrasting these two ages can serve as a model for multiple hypothesis-driven investigations for the mechanisms underlying refractoriness at P7. This approach has identified novel small-molecule drug candidates for refractory neonatal seizures.

References

1. Massey SL, Jensen FE, Abend NS (2018) Electroencephalographic monitoring for seizure identification and prognosis in term neonates. Semin Fetal Neonatal Med 23:168–174

2. Boylan GB, Stevenson NJ, Vanhatalo S (2013) Monitoring neonatal seizures. Semin Fetal Neonatal Med 18:202–208

3. Stevenson NJ, Boylan GB, Hellström-Westas L, Vanhatalo S (2016) Treatment trials for neonatal seizures: the effect of design on sample size. PLoS One 11:e0165693

4. Scher MS, Alvin J, Gaus L, Minnigh B, Painter MJ (2003) Uncoupling of EEG-clinical neonatal seizures after antiepileptic drug use. Pediatr Neurol 28:277–280

5. Pressler R, Lagae L (2019) Why we urgently need improved seizure and epilepsy therapies for children and neonates. Neuropharmacology 170:107854. https://doi.org/10.1016/j.neuropharm.2019.107854

6. Kwon JM et al (2011) Clinical seizures in neonatal hypoxic-ischemic encephalopathy have no independent impact on neurodevelopmental outcome: secondary analyses of data from the neonatal research network hypothermia trial. J Child Neurol 26:322–328

7. Soul JS et al (2019) Recommendations for the design of therapeutic trials for neonatal seizures. Pediatr Res 85:943–954

8. Kang SK, Kadam SD (2015) Neonatal seizures: impact on neurodevelopmental outcomes. Front Pediatr 3:101

9. Galanopoulou AS, Pitkänen A, Buckmaster PS, Moshé SL (2017) Chapter 75 – what do models model? What needs to be modeled? In: Pitkänen A, Buckmaster PS, Galanopoulou AS, Moshé SL (eds) Models of seizures and epilepsy, 2nd edn. Academic, London, pp 1107–1119. https://doi.org/10.1016/B978-0-12-804066-9.00077-8

10. Kang SK, Kadam SD (2014) Pre-clinical models of acquired neonatal seizures: differential effects of injury on function of chloride

co-transporters. Austin J Cerebrovasc Dis Stroke 1:1026

11. Kang SK, Markowitz GJ, Kim ST, Johnston MV, Kadam SD (2015) Age- and sex-dependent susceptibility to phenobarbital-resistant neonatal seizures: role of chloride co-transporters. Front Cell Neurosci 9:173

12. Kang SK, Ammanuel S, Adler DA, Kadam SD (2020) Rescue of PB-resistant neonatal seizures with single-dose of small-molecule TrkB antagonist show long-term benefits. Epilepsy Res 159:106249

13. Kharod SC, Carter BM, Kadam SD (2018) Pharmaco-resistant neonatal seizures: critical mechanistic insights from a chemoconvulsant model. Dev Neurobiol 78:1117. https://doi.org/10.1002/dneu.22634

14. Pressler RM et al (2015) Bumetanide for the treatment of seizures in newborn babies with hypoxic ischaemic encephalopathy (NEMO): an open-label, dose finding, and feasibility phase 1/2 trial. Lancet Neurol 14:469–477

15. Doyon N, Vinay L, Prescott SA, De Koninck Y (2016) Chloride regulation: a dynamic equilibrium crucial for synaptic inhibition. Neuron 89:1157–1172

16. Kaila K, Price TJ, Payne JA, Puskarjov M, Voipio J (2014) Cation-chloride cotransporters in neuronal development, plasticity and disease. Nat Rev Neurosci 15:637

17. Spoljaric A et al (2017) Vasopressin excites interneurons to suppress hippocampal network activity across a broad span of brain maturity at birth. Proc Natl Acad Sci 114:E10819–E10828

18. Noebels J (2015) Pathway-driven discovery of epilepsy genes. Nat Neurosci 18:344–350

19. Smith RS, Walsh CA (2020) Ion channel functions in early brain development. Trends Neurosci 43:103–114

20. Kadam SD, Sullivan BJ, Goyal A, Blue ME, Smith-Hicks C (2019) Rett syndrome and CDKL5 deficiency disorder: from bench to clinic. Int J Mol Sci 20:5098

21. Johnston MV et al (2014) Twenty-four hour quantitative-EEG and in-vivo glutamate biosensor detects activity and circadian rhythm dependent biomarkers of pathogenesis in Mecp2 null mice. Front Syst Neurosci 8:118

22. Olson HE et al (2019) Cyclin-dependent kinase-like 5 deficiency disorder: clinical review. Pediatr Neurol 97:18–25

23. Sullivan BJ et al (2020) Low-dose Perampanel rescues cortical gamma dysregulation associated with parvalbumin interneuron GluA2 upregulation in epileptic Syngap1+/− mice. Biol Psychiatry 87:829. https://doi.org/10.1016/j.biopsych.2019.12.025

24. Comi AM, Johnston MV, Wilson MA (2005) Immature mouse unilateral carotid ligation model of stroke. J Child Neurol 20:980–983

25. Kadam SD et al (2017) Methodological standards and interpretation of video-electroencephalography in adult control rodents. A TASK1-WG1 report of the AES/ILAE Translational Task Force of the ILAE. Epilepsia 58(Suppl 4):10–27

26. Baram TZ et al (2012) Fragmentation and unpredictability of early-life experience in mental disorders. Am J Psychiatry 169:907–915

27. Weiergräber M, Papazoglou A, Broich K, Müller R (2016) Sampling rate, signal bandwidth and related pitfalls in EEG analysis. J Neurosci Methods 268:53–55

28. Moyer JT et al (2017) Standards for data acquisition and software-based analysis of in vivo electroencephalography recordings from animals. A TASK1-WG5 report of the AES/ILAE Translational Task Force of the ILAE. Epilepsia 58:53–67

29. Maynard D, Prior PF, Scott DF (1969) Device for continuous monitoring of cerebral activity in resuscitated patients. Br Med J 4:545–546

30. Boylan GB, Kharoshankaya L, Mathieson SR (2019) Chapter 18 – diagnosis of seizures and encephalopathy using conventional EEG and amplitude integrated EEG. In: de Vries LS, Glass HC (eds) Handbook of clinical neurology, vol 162. Elsevier, Amsterdam, pp 363–400

31. Shellhaas RA et al (2011) The American Clinical Neurophysiology Society's guideline on continuous electroencephalography monitoring in neonates. J Clin Neurophysiol Off Publ Am Electroencephalogr Soc 28:611–617

32. Temko A, Marnane W, Boylan G, Lightbody G (2015) Clinical implementation of a neonatal seizure detection algorithm. Decis Support Syst 70:86–96

33. Wilcox KS, West PJ, Metcalf CS (2020) The current approach of the Epilepsy Therapy Screening Program contract site for identifying improved therapies for the treatment of pharmacoresistant seizures in epilepsy. Neuropharmacology 166:107811

34. Carter BM, Sullivan BJ, Landers JR, Kadam SD (2018) Dose-dependent reversal of KCC2 hypofunction and phenobarbital-resistant neonatal seizures by ANA12. Sci Rep 8:11987

35. Kang SK, Johnston MV, Kadam SD (2015) Acute TrkB inhibition rescues phenobarbital-resistant seizures in a mouse model of neonatal ischemia. Eur J Neurosci 42:2792–2804

36. Kang SK et al (2018) Sleep dysfunction following neonatal ischemic seizures are differential by neonatal age of insult as determined by

qEEG in a mouse model. Neurobiol Dis 116:1–12

37. Kipnis PA, Sullivan BJ, Kadam SD (2019) Sex-dependent signaling pathways underlying seizure susceptibility and the role of chloride cotransporters. Cell 8:448

38. Khirug S et al (2010) A single seizure episode leads to rapid functional activation of KCC2 in the neonatal rat hippocampus. J Neurosci 30:12028

39. Gilad D et al (2015) Homeostatic regulation of KCC2 activity by the zinc receptor mZnR/GPR39 during seizures. Neurobiol Dis 81:4–13

40. Puskarjov M et al (2015) BDNF is required for seizure-induced but not developmental up-regulation of KCC2 in the neonatal hippocampus. Neuropharmacology 88:103–109

41. Gale K (1995) Chemoconvulsant seizures: advantages of focally-evoked seizure models. Ital J Neurol Sci 16:17–25

42. Williams PA et al (2009) Development of spontaneous recurrent seizures after kainate-induced status epilepticus. J Neurosci 29:2103–2112

43. Rice JE, Vannucci RC, Brierley JB (1981) The influence of immaturity on hypoxic-ischemic brain damage in the rat. Ann Neurol 9:131–141

44. Burnsed J, Skwarzyńska D, Wagley PK, Isbell L, Kapur J (2019) Neuronal circuit activity during neonatal hypoxic–ischemic seizures in mice. Ann Neurol 86:927–938

45. Kadam SD, White AM, Staley KJ, Dudek FE (2010) Continuous electroencephalographic monitoring with radio-telemetry in a rat model of perinatal hypoxia–ischemia reveals progressive post-stroke epilepsy. J Neurosci 30:404–415

46. Adami RR et al (2016) Distinguishing arterial ischemic stroke from hypoxic-ischemic encephalopathy in the neonate at birth. Obstet Gynecol 128:704–712

Part V

Invertebrate/Non-Mammalian Models

Chapter 12

Methods to Investigate Seizures and Associated Cognitive Decline Using Zebrafish Model

Brandon Kar Meng Choo and Mohd. Farooq Shaikh

Abstract

Epileptic seizures are the manifestation of several signs and/or symptoms due to heightened brain activity, causing a variety of disturbances such as motor and cognition dysfunction. In contrast, epilepsy is characterized by a chronic tendency to spawn these epileptic seizures. One way of studying epilepsy or seizures is to induce either acute seizures or epilepsy in animals via chemoconvulsants such as pentylenetetrazol (acute seizures) or kainic acid (chronic seizures), though pentylenetetrazol can also produce chronic seizures via kindling. Zebrafish are of great utility as an animal model as they have a high breeding rate and are genetically similar to humans. By intraperitoneally injecting chemoconvulsants into zebrafish, they develop seizure-like behavior, which can be scored to determine severity. By recording this behavior, the resulting locomotion pattern and parameters can also be tracked via software analysis. To investigate the cognitive decline comorbidity of epileptic seizures, mazes such as the T-maze and three-axis maze can be used to evaluate the learning and memory ability of the zebrafish via operant conditioning. This is done by comparing and contrasting the time taken for a zebrafish to reach the location of a reward over successive trials.

Key words Epilepsy, Seizures, Zebrafish (*Danio rerio*), Kindling, T-maze, Three-axis maze, Cognitive decline

1 Introduction

Epileptic seizures are commonly termed as a short-term appearance of many signs and/or symptoms due to extraordinarily excessive or synchronized brain activity. Generally, seizures may disturb autonomic, sensory, and motor function, as well as cognition, memory, emotional state, or behavior, though not necessarily simultaneously. In contrast, epilepsy is an assortment of neurological disorders that are characterized by an enduring predisposition toward producing epileptic seizures. The International League Against Epilepsy (ILAE) in 2014 defined epilepsy to be a brain disease which meets any one of the three conditions. The first condition is having more than one unprovoked or reflex seizures which occur

Divya Vohora (ed.), *Experimental and Translational Methods to Screen Drugs Effective Against Seizures and Epilepsy*, Neuromethods, vol. 167, https://doi.org/10.1007/978-1-0716-1254-5_12,

more than 24 h apart. The second condition is the occurrence of an unprovoked or reflex seizure with the possibility of successive seizures being similar to the typical relapse risk of at least 60%, after having two unprovoked seizures which occur within a decade. The final condition is the diagnosis of epilepsy syndrome [1]. While the fundamental cause of epilepsy is not always clear, anti-convulsant drugs or anti-epileptic drugs (AEDs), as they are frequently termed, can be used for the symptomatic treatment of epilepsy. While AED side effects are typically patient dependent, its severity also depends on the dose and is aggravated by the simultaneous use of several AEDs. This is detrimental as multiple AEDs are routinely prescribed together at high doses in an effort to stop seizures entirely [2]. Cognitive impairment is possibly one of the most noteworthy adverse effects of AEDs, though the degree of impairment varies between AEDs. This is ironic as cognitive impairment is also one of the comorbidities of epilepsy itself [3].

Among the early stages in the development of a novel treatment is animal testing. But before developing animal models of epilepsy, a method of inducing seizures in animals reliably must first be determined. One of the most frequently used methods of inducing seizures in animals is via chemoconvulsants. An example of a chemoconvulsant among the many different available is pentylenetetrazol (PTZ). Different animal species react differently when exposed to PTZ, such as how adult zebrafish progress through a characteristic set of movements and postures which end with generalized tonic-clonic seizures [4]. PTZ is thought to induce seizures predominantly by binding to the γ-aminobutyric acid (GABA$_A$) receptor and impeding the neuroinhibitory action of GABA [4]. Chemoconvulsants may be utilized to produce acute and chronic seizures, although limitations in cost and time often limit studies to acute seizures.

While the PTZ model of acute seizures is a well-accepted animal model, it is not a true model of epilepsy. This is because one important characteristic of epilepsy is the enduring predisposition to spawn epileptic seizures [5], whereas the PTZ-induced acute seizures diminish after a short period of time. Nonetheless, it is a convenient model for studying molecular mechanisms as compared to chronic epilepsy models which better mimic the clinical epileptic condition [6] but are time-consuming by design. However, it is also possible to develop chronic seizures using PTZ by using a process known as kindling, which is a term used to describe the initiation of full-blown seizures after repeated, intermittent exposure to a stimulus at a level below the typical threshold for inducing full-blown seizures [7]. Other than PTZ kindling, other chemoconvulsants such as kainic acid can also be used to produce chronic epilepsy in zebrafish by acting on the specific kainate receptors in the central nervous system [8] to produce continuous seizures and neuronal toxicity [9–11]. In addition, kindling is a phenomenon where a sub-convulsive stimulus (either chemical or electrical), if applied repetitively and intermittently, will ultimately lead to the generation of full-blown convulsions [12]. The readers may refer to Chapters 5, 6 and 7 for detail description of these models.

Once the animal model of epilepsy and the method of inducing epilepsy are ascertained, a technique for assessing the degree of seizure activity and possibly comorbidities is needed. Adult zebrafish are employed to model other phenotypes such as the more complicated behaviors and biomarkers usually linked to epilepsy due to their high breeding rate and genetic similarity to humans [13]. Adult zebrafish usually undergo testing inside a tank in which they can be observed so that their seizure behavior can be scored according to a predefined scoring system. The scoring may be done manually in real time or afterward via video recording in combination with automatic software-assisted analysis of swimming pattern and parameters. The side points of view for the observation tank can be used for the neurophenotypic classification of the responses which result in adult zebrafish treated with a chemoconvulsant, as they are very similar to those observed during a seizure. While the abnormal response displayed by the zebrafish will vary based on the chemoconvulsant used, the conventional endpoints which are used include rapid twitching; loss of body posture; hyperactive, spiral, or circular swimming; paralysis or immobility; body contractions similar to spasms; and death [14].

Cognitive dysfunction is said to be a decline in one of the six cognitive domains as defined in the *Diagnostic and Statistical Manual of Mental Disorders, Fifth Edition* (DSM-5). These domains are complex attention, executive function, learning and memory, language, perceptual-motor, and social cognition [15]. The domain of learning and memory consists of both immediate memory and recent memory (including free and cued recall as well as recognition memory) [16]. Spatial memory can be assessed in zebrafish using the T-maze, which relies on operant learning that encourages the zebrafish to favor entering a certain compartment in the T-maze for a reward [17]. Another test of learning and memory is the three-axis maze, which assesses egocentric memory by requiring zebrafish to navigate a route across the X, Y, and Z axes to reach an expected food source, independently of environmental cues [11].

2 Materials

2.1 Animals

This procedure uses adult zebrafish (*Danio rerio*), which is typically defined as about 3 months of age or older. The average weight of these zebrafish is typically between 0.4 and 0.6 g. All zebrafish should be held under standard husbandry conditions. This involves maintaining the zebrafish tanks at a water temperature of between 26 and 30 °C, at a water pH of between pH 6.8 and 7.1, and under a constant light intensity of approximately 250 lux, with a 12 h light-dark cycle. The zebrafish should be fed two to three times a day with dry fish food, and their diet may be supplemented with live brine shrimps (*Artemia*). The tanks should also be equipped with a water circulation system to provide constant aeration. The recommended housing density is approximately one zebrafish per liter of tank capacity, and males and females should be kept separate.

2.2 Drugs

1. Pentylenetetrazol (PTZ) from Sigma-Aldrich (United States).
2. Kainic acid (KA) from Sigma-Aldrich (United States).
3. Benzocaine from Sigma-Aldrich (United States).
4. Normal saline solution.

2.3 Pre-working Preparations

Before the experiment, freshly prepare the appropriate pro-convulsant in normal saline solution as per the concentrations below. Keep the solutions covered from light and on ice when not in use:

Pentylenetetrazol (acute)—220 g/L.

Pentylenetetrazol kindling (chronic)—80 g/L.

Kainic acid (chronic)—3 g/L.

Also, prepare a stock solution of the anesthetic benzocaine in ethanol at a concentration of 30 mg/L. Store the stock solution at room temperature and tightly cap. Prior to the experiment, freshly prepare a working solution of benzocaine by diluting 1 mL of the stock benzocaine solution in 500 mL of the same water used for the adult zebrafish to obtain a final concentration of 0.06 mg/L. Keep at room temperature, and cover the top of the solution container when not being used for extended periods to reduce evaporation.

2.4 Equipment

1. Tripod.
2. Tripod-mountable camcorder or digital camera.
3. Fish recording tank made with transparent acrylic on one of the long sides and opaque white acrylic on the remaining three sides.
4. 10 µL Hamilton syringe (700 series, Hamilton 80400).
5. 30 G needles (BD PrecisionGlide™).
6. Makeup sponges.
7. Digital weighing scale with an accuracy of $\pm$ 0.1 g.

2.5 Setup

Fast the adult zebrafish for 24 h prior to the experiment. Ensure that the room in which both the adult zebrafish and the behavioral recording setup are kept is maintained at a temperature of between

26 and 30 °C, with the humidity maintained at between 50% and 60%. The zebrafish should be left in the room to acclimatize at least 2 h before the start of the experiment to discourage a novel tank response from the zebrafish. The behavioral recording setup involves first placing the fish recording tank in a relatively quiet area and away from any bright sources of light that cause glare visible to the camera. Next, fill up the tank three-quarters of the way with the water that the fish are normally kept in, and wipe off any water droplets on the transparent side. After that, set up the tripod in front of the clear side of the tank and mount the camcorder on it.

Prepare the intraperitoneal injection setup by making an incision 30–40 mm across and 10–15 mm deep in the makeup sponge for the purpose of restraining and holding the zebrafish during the intraperitoneal injection process. After that, saturate the sponge with water and place it in a 60 mm × 15 mm petri dish.

2.6 Adult Zebrafish Tests of Cognitive Function

2.6.1 The Zebrafish T-Maze Cognitive Test (Acute Seizures)

The zebrafish T-maze (Fig. 1) consists of one long (45.72 cm) arm with a divider at one end to form a small starting chamber and two short (30.48 cm) arms connected to the long arm. One short arm is connected to a deeper square chamber (22.86 cm × 22.86 cm). The entire T-maze is made from opaque white acrylic to minimize external visual cues.

The T-maze behavior test should be conducted in a room which is kept at a temperature of between 26 and 30 °C and humidity between 50% and 60%. The T-maze should be filled with water to approximately three-quarters full. After that, make sure that there are no bright reflections reflected by the water in the maze and that the lighting of the T-maze is fairly uniform to minimize shadows and external visual cues. Next, mount a camcorder above the T-maze such that the entire maze is within the camcorder's field of view.

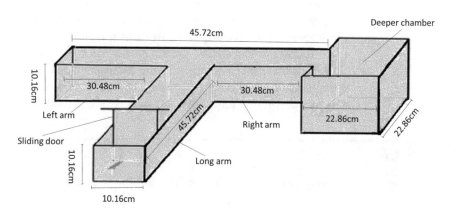

Fig. 1 Diagrammatic representation of a zebrafish T-maze with a zebrafish in the starting chamber and colored blue to represent the maze being filled with water

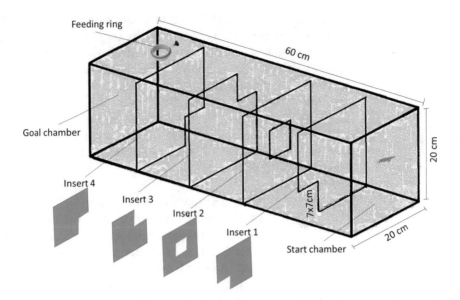

Fig. 2 Diagrammatic representation of the zebrafish three-axis maze with a zebrafish in the starting chamber and colored blue to represent the maze being filled with water. Note the four inserts that divide the tank into separate chambers and the floating feeding ring in the goal chamber

2.6.2 The Zebrafish Three-Axis Maze Cognitive Test (Chronic Seizures)

The zebrafish three-axis maze (Fig. 2) consists of a 60-cm-long, 20-cm-wide, and 20-cm-tall tank made from opaque white acrylic on three sides to minimize external visual cues. One of the remaining long sides is made from transparent acrylic. The tank is divided equally lengthwise into five sections by inserts that are 20 cm wide and 20 cm tall. Each insert has a 7-cm-tall and 7-cm-wide hole cut into the insert at different locations.

The three-axis maze test should also be conducted in a room which is kept at a temperature of between 26 and 30 °C and humidity between 50% and 60%. The three-axis maze should be filled with water to approximately three-quarters full. Next, a fish feeding ring should be attached to one of the opaque walls of the goal chamber just above the surface of the water such that it floats. After that, set up the tripod in front of the clear side of the maze and mount the camcorder on it. Ensure that the lighting of the three-axis maze is fairly uniform to minimize shadows and external visual cues. Also ensure that the recording area is relatively quiet and that there is no glare from the tank that is visible to the camera.

3 Methodology

1. Transfer the zebrafish from a holding tank into the benzocaine anesthetic solution.

2. Allow the zebrafish to remain in the benzocaine until the zebrafish loses balance (turns upside down) and stops swimming.

3.1 Adult Zebrafish Intraperitoneal Injection and Behavioral Recording (Acute Model)

3. Remove the zebrafish from the benzocaine, and place it on the weighing scale to determine its weight.

4. Place the zebrafish inside the incision made in the sponge, with the belly side facing upward.

8. Draw 10μL of pro-convulsant solution per gram of zebrafish body weight (*see* **Notes 1** and **2**).

5. Insert the needle of the syringe into the midline between the pelvic fins to inject the pro-convulsant intraperitoneally.

6. Transfer the zebrafish to the fish recording tank, and begin recording for 10 min once the zebrafish fully recovers from anesthesia and begins swimming normally (*see* **Note 3**).

3.2 Adult Zebrafish T-Maze Test of Cognitive Function (Acute)

1. After the behavioral recording, transfer the zebrafish to the starting chamber, and allow the zebrafish to calm down and begin swimming at a normal pace.

2. Start recording and lift the divider as quickly as possible without alarming the zebrafish (*see* **Notes 4–6**).

3. Continue recording for 5 min or until the zebrafish reaches the deeper chamber and stays there for 20–30 s.

4. Chase any zebrafish which do not reach the deeper chamber after 5 min using a net, and allow them to remain there for a few seconds.

5. Transfer the zebrafish back into a holding tank.

6. Repeat the entire procedure 3 and 24 h after the initial T-maze trial.

3.3 Adult Zebrafish Three-Axis Maze Test of Cognitive Function (Chronic)

The procedure for investigating seizures and associated cognitive decline using a chronic model is similar to the acute model described above for behavioral recording, with some modifications:

1. In the case of PTZ kindling, the PTZ must be intraperitoneally injected into the zebrafish daily.

2. Kainic acid is intraperitoneally injected only once.

3. After the intraperitoneal injection, the behavioral recording is repeated once daily.

4. After the behavioral recording, the three-axis maze test is also repeated once daily.

However, cognitive function for chronic seizure zebrafish models is assessed using the three-axis maze rather than the T-maze. This test is divided into two major parts:

3.3.1 Initial Training

1. Fast the adult zebrafish for 24 h prior to the experiment, and there should be no subsequent feeding of the zebrafish outside the experiment.

2. Place some fish food into the feeding ring (*see* **Note 7**).

3. Remove the inserts from the maze if inserted.

4. Place a zebrafish into the starting chamber, and allow it to swim until it reaches the goal chamber and has fed on the fish food for about 30 s.

5. Repeat the training procedure after a short period of rest.

3.3.2 Cognitive Function Testing

1. Place the inserts into the three-axis maze in any order but in a fixed order for a given study.

2. Transfer the zebrafish to the starting chamber, and allow it to swim freely for 10 min or 30 s after it begins feeding at the goal chamber to establish a baseline.

3. After the behavioral recording, repeat the three-axis maze test.

4. Transfer the zebrafish back into a holding tank.

5. Repeat the test once daily.

4 Data Acquisition and Analysis

The Smart V3.0.05 tracking software (Panlab, Harvard Apparatus) can be used for the automated tracking of zebrafish swimming patterns and locomotion parameters (time spent in either half of the tank, distance travelled in either half of the tank, and others as desired) using the previously recorded videos. Transfer latencies or the time taken for the zebrafish to reach a particular chamber can also be determined via the Smart software. In the case of seizure scoring, the video is simply played back, and the seizure score of the zebrafish is manually determined according to the seizure scores given below.

4.1 The PTZ Seizure Model

A dose of between 220 and 250 mg/kg of PTZ is sufficient to produce full-blown seizures in zebrafish [18], with 170 mg/kg limiting the resulting seizure scores to a maximum of 4 [19] and 80 mg/kg being a sub-convulsive dose for kindling purposes [7]. The highest seizure score over the course of each 10 min video is taken to be the seizure score for that zebrafish. Another parameter is seizure onset time, which is the time taken for the seizures to reach a predetermined stage (typically scores 4–6 as they are considered to be more severe). The zebrafish seizure scoring system for PTZ-induced acute seizures is a slight modification from the scoring system originally defined by Mussulini et al. [20] and used by Kundap et al. [19] and is given below:

Score 1: Short swim mainly at the bottom of the tank.

Score 2: Increased swimming activity and high frequency of opercular movement.

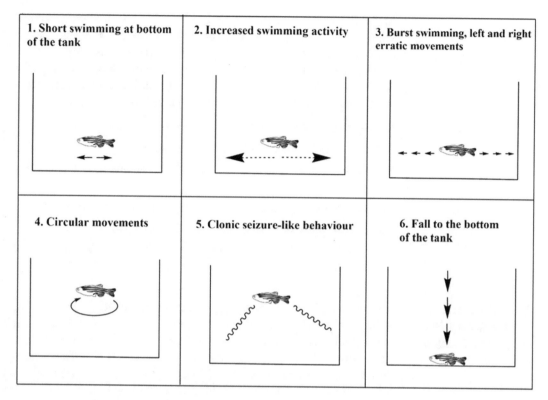

Fig. 3 Diagrammatic representation of pentylenetetrazol-induced seizure patterns denoted by seizure scores

Score 3: Burst swimming, left and right movements, as well as erratic movements.

Score 4: Circular movements.

Score 5: Clonic seizure-like behavior (abnormal whole-body rhythmic muscular contraction).

Score 6: Fall to the bottom of the tank, tonic seizure-like behavior (sinking to the bottom of the tank, loss of body posture, and principally rigid body extension) (Fig. 3).

4.2 Adult Zebrafish T-Maze Test of Cognitive Function (Acute)

The T-maze is designed to measure transfer latency (TL), which is the time taken for the zebrafish to reach the deeper chamber, and this measured 0, 3 and 24 h after the PTZ challenge. The transfer latencies are then expressed as inflection ratios (IR) whereby the inflexion ratio after 3 h $(IR_3) = (TL_0 - TL_3)/(TL_3)$ and the inflexion ratio after 24 h $(IR_{24}) = (TL_0 - TL_{24})/(TL_{24})$. A higher ratio represents an improvement in cognitive function, and a lower ratio represents an impairment.

4.3 The Kainic Acid Seizure Model

A single dose of KA dose of 3 mg/kg produces seizures in adult zebrafish that diminish over the course of several days [7]. The highest seizure score over the course of each 10 min video is taken to be the seizure score for that zebrafish. Another parameter is seizure onset time, which is the time taken for the seizures to reach a predetermined stage (typically scores 4–6 as they are considered to be more severe). The zebrafish seizure scoring for KA-induced chronic seizures is a slight modification from the scoring system defined by Alfaro et al. [9] and is given below:

Score 1: Rigidity and hyperventilation of the animal.

Score 2: Whirlpool-like swimming behavior.

Score 3: Rapid muscular uncontrolled movements from right to left.

Score 4: Abnormal and spasmodic muscular contractions.

Score 5: Rapid whole-body clonus-like convulsions.

Score 6: Sinking to the bottom of the tank and spasms for several minutes.

4.4 Adult Zebrafish Three-Axis Maze Test of Cognitive Function (Chronic)

The three-axis maze is design to measure transfer latency (TL), which is the time taken for the zebrafish to reach the feeding ring in the goal chamber and in each daily trial. The transfer latencies are then expressed as inflection ratios (IR) whereby the inflexion ratio on day "x" is given by $(IR_x) = (TL_0 - TL_x)/(TL_x)$, whereby TL_0 represents the transfer latency for the initial baseline recording. A higher ratio represents an improvement in cognitive function, and a lower ratio represents an impairment.

5 Notes

1. Lighter zebrafish require smaller intraperitoneal injection volumes, which could have accuracy implications. Thus, the use of very light zebrafish (~0.3 g or less) should be avoided.

2. In contrast, heavier zebrafish require larger intraperitoneal injection volumes, which may pose problems for those with shorter digits. A 25 µL Hamilton syringe could also be used in this case.

3. If the zebrafish do not recover within 1–2 min, dilute the benzocaine to about 550 or 600 mL.

4. When monitoring the T-maze, be aware that looking over the maze from too close can cause your shadow to be cast on the maze and interfering with subsequent software tracking of the locomotion pattern. Instead, remotely monitor the maze if possible or stand further away.

5. Also, do not lift the divider slowly enough that the zebrafish slips under the opening before the divider is fully lifted as this will make it difficult for the software to track the initial movements due to the shadow that the divider casts.

6. In contrast, lifting the divider too quickly may startle the zebrafish, causing it to swim faster than it would normally, thus potentially altering the time taken to reach the deeper chamber in an erroneous manner. Fast swimming by the zebrafish may also create ripples which cause reflections that interfere with software tracking of the locomotion pattern.

7. After a few trials, remove any sunken food or zebrafish excrement, and replenish the food in the feeding ring.

References

1. Fisher RS, Acevedo C, Arzimanoglou A, Bogacz A, Cross JH, Elger CE, Engel J, Forsgren L, French JA, Glynn M, Hesdorffer DC, Lee BI, Mathern GW, Moshé SL, Perucca E, Scheffer IE, Tomson T, Watanabe M, Wiebe S (2014) ILAE official report: a practical clinical definition of epilepsy. Epilepsia 55(4):475–482. https://doi.org/10.1111/epi.12550

2. Zaccara G, Perucca P (2015) Prevention and management of side-effects of antiepileptic drugs. In: The treatment of epilepsy. Wiley, Hoboken, pp 275–287. https://doi.org/10.1002/9781118936979.ch20

3. Holmes GL (2015) Cognitive impairment in epilepsy: the role of network abnormalities. Epileptic Disord 17(2):101–116. https://doi.org/10.1684/epd.2015.0739

4. Berghmans S, Hunt J, Roach A, Goldsmith P (2007) Zebrafish offer the potential for a primary screen to identify a wide variety of potential anticonvulsants. Epilepsy Res 75(1):18–28. https://doi.org/10.1016/j.eplepsyres.2007.03.015

5. Fisher RS, Boas WVE, Blume W, Elger C, Genton P, Lee P, Engel J Jr (2005) Epileptic seizures and epilepsy: definitions proposed by the International League Against Epilepsy (ILAE) and the International Bureau for Epilepsy (IBE). Epilepsia 46(4):470–472

6. Curia G, Longo D, Biagini G, Jones RSG, Avoli M (2008) The pilocarpine model of temporal lobe epilepsy. J Neurosci Methods 172 (2):143–157. https://doi.org/10.1016/j.jneumeth.2008.04.019

7. Kundap UP, Paudel YN, Kumari Y, Othman I, Shaikh MF (2019) Embelin prevents seizure and associated cognitive impairments in a pentylenetetrazole-induced kindling zebrafish model. Front Pharmacol 10(315). https://doi.org/10.3389/fphar.2019.00315

8. Zheng X-Y, Zhang H-L, Luo Q, Zhu J (2010) Kainic acid-induced neurodegenerative model: potentials and limitations. Biomed Res Int 2011:457079

9. Alfaro JM, Ripoll-Gómez J, Burgos JS (2011) Kainate administered to adult zebrafish causes seizures similar to those in rodent models. Eur J Neurosci 33(7):1252–1255

10. Menezes FP, Rico EP, Da Silva RS (2014) Tolerance to seizure induced by kainic acid is produced in a specific period of zebrafish development. Prog Neuro-Psychopharmacol Biol Psychiatry 55:109–112

11. Mussulini BHM, Vizuete AFK, Braga M, Moro L, Baggio S, Santos E, Lazzarotto G, Zenki KC, Pettenuzzo L, da Rocha JBT (2018) Forebrain glutamate uptake and behavioral parameters are altered in adult zebrafish after the induction of Status Epilepticus by kainic acid. Neurotoxicology 67:305–312

12. Dhir A (2012) Pentylenetetrazol (PTZ) kindling model of epilepsy. Curr Protoc Neurosci 58(1):9.37.31–39.37.12

13. Wong K, Stewart A, Gilder T, Wu N, Frank K, Gaikwad S, Suciu C, DiLeo J, Utterback E, Chang K, Grossman L, Cachat J, Kalueff AV (2010) Modeling seizure-related behavioral and endocrine phenotypes in adult zebrafish. Brain Res 1348:209–215. https://doi.org/10.1016/j.brainres.2010.06.012

14. Stewart AM, Desmond D, Kyzar E, Gaikwad S, Roth A, Riehl R, Collins C, Monnig L, Green J, Kalueff AV (2012) Perspectives of zebrafish models of epilepsy: what, how and where next? Brain Res Bull 87(2):135–143

15. American Psychiatric Association (2013) Diagnostic and statistical manual of mental disorders (DSM-5®). American Psychiatric Pub

16. Goodkin K, Fernandez F, Forstein M, Miller EN, Becker JT, Douaihy A, Cubano L, Santos FH, Filho NS, Zirulnik J, Singh D (2011) A perspective on the proposal for neurocognitive disorder criteria in DSM-5 as applied to HIV-associated neurocognitive disorders.

Neuropsychiatry 1(5):431–440. https://doi.org/10.2217/npy.11.57

17. Bailey JM, Oliveri AN, Levin ED (2015) Pharmacological analyses of learning and memory in zebrafish (Danio rerio). Pharmacol Biochem Behav 139(Pt B):103–111. https://doi.org/10.1016/j.pbb.2015.03.006

18. Banote RK, Koutarapu S, Chennubhotla KS, Chatti K, Kulkarni P (2013) Oral gabapentin suppresses pentylenetetrazole-induced seizure-like behavior and cephalic field potential in adult zebrafish. Epilepsy Behav 27(1):212–219

19. Kundap UP, Kumari Y, Othman I, Shaikh M (2017) Zebrafish as a model for epilepsy-induced cognitive dysfunction: a pharmacological, biochemical and behavioral approach. Front Pharmacol 8:515

20. Mussulini BHM, Leite CE, Zenki KC, Moro L, Baggio S, Rico EP, Rosemberg DB, Dias RD, Souza TM, Calcagnotto ME, Campos MM, Battastini AM, de Oliveira DL (2013) Seizures induced by pentylenetetrazole in the adult zebrafish: a detailed behavioral characterization. PLoS One 8(1):e54515. https://doi.org/10.1371/journal.pone.0054515

Chapter 13

C. elegans as a Potential Model for Acute Seizure-Like Activity

Alistair Jones, Anthony G. Marson, Vincent T. Cunliffe, Graeme J. Sills, and Alan Morgan

Abstract

Conventional rodent models have provided invaluable tools for the discovery and characterization of antiepileptic drugs (AEDs). Nevertheless, around one third of people with epilepsy do not respond to currently available treatments, and so new AEDs are surely needed. However, traditional rodent models are expensive, pose ethical problems, and are poorly suited to high-throughput drug discovery. As a result, methods for front-line screening of compound libraries in simpler seizure models such as fruit flies and zebrafish have been successfully developed in recent years. Here we describe an assay of seizure-like activity using the nematode worm, *Caenorhabditis elegans*. Our main aim here is to explain how to perform the method, highlighting advantages and limitations of both assay and the animal model. In addition, we give examples of how the assay can be used to determine pathogenicity in epilepsy-associated genetic variants, to screen for anticonvulsants, and to identify AED mechanisms of action.

Key words Seizure assay, Nematode, Pentylenetetrazole, Drug screening, High-throughput

1 Introduction

Identification of antiepileptic drugs (AEDs) predominantly relies on mouse or rat paradigms. These animals have historically provided a great wealth of information for many of the AEDs used today [1–5]. During early phases of drug profiling, rodent acute seizure models offer valuable insight into new therapeutics. For example, the data collected can be used to make informed choices concerning more specialized seizure or chronic epilepsy models to pursue in the future [6]. However, the expense, ethical concerns, and potential species-specific effects of compounds in these experiments may hamper translating positive treatments to humans [1]. Therefore, incorporating simpler model organisms into drug screening pipelines can offer a cheaper, faster, and ethical alternative, while providing a second screening organism in early

Divya Vohora (ed.), *Experimental and Translational Methods to Screen Drugs Effective Against Seizures and Epilepsy*, Neuromethods, vol. 167, https://doi.org/10.1007/978-1-0716-1254-5_13,
© Springer Science+Business Media, LLC, part of Springer Nature 2021

compound profiling [2, 7]. The fruit fly, *Drosophila melanogaster*, and the zebrafish, *Danio rerio*, are well-established animal models that have been successfully used for high-throughput anticonvulsant drug screening [8, 9]. In contrast, although the nematode worm *Caenorhabditis elegans* has been extensively used to model a wide variety of neurological disorders [10], there are relatively few studies where it has been used as a model organism for the study of epilepsy [11–19].

C. elegans was established as an animal model in the 1960s, largely through the work of Nobel laureate Sydney Brenner, bringing the worm to the forefront as an in vivo platform to study biological processes such as development and neuronal function [20]. Since this point, researchers have made this simple organism arguably the most thoroughly understood animal model in use today. For instance, *C. elegans* is unique among animals in that the position and developmental fate of every cell from zygote to adult is fully documented [21, 22], as are all of the approximately 6400 synaptic connections made by its 302 neurons [23]. These neuronal synapses are architecturally similar to humans, and the majority of neuronal proteins involved in synaptic release are highly conserved, as are the neurotransmitters released [24]. Indeed, many of the proteins involved in neurotransmission were originally discovered using genetic screens in *C. elegans* [24].

Recent genomic studies have revealed that genetic epilepsies are caused by mutations in multiple genes [25–27]. As around 40% of human disease-related genes have *C. elegans* orthologues [28], this presents an opportunity to model genetic epilepsies in worms. The powerful and facile genetics of *C. elegans* allows incorporation of multiple mutations into the nematodes in a rapid and cost-effective manner. These modified animals could then potentially be utilized in high-throughput compound screens against tailored transgenic backgrounds. As *C. elegans* exist predominantly as self-fertilizing hermaphrodites, there is no need to maintain mutant populations by mating, as is the case for all other popular animal models. Nevertheless, rare males can be isolated and mated with hermaphrodites, which enables genetic crossing for applications such as construction of double mutant strains. Incorporating mutations is comparatively easy in *C. elegans*, due to well-established genetic manipulation techniques. New experimental DNA of interest can be introduced via microinjection into the worm gonad, a versatile process allowing addition of large extrachromosomal arrays for over-expression studies or more subtle CRISPR-based approaches [29, 30].

C. elegans was the first multicellular organism to have its genome completely sequenced, which has been reproduced in "WormBase," a user-friendly online database with highly accurate annotations. Worm strains harboring multiple alternative mutant alleles can be identified for most genes via WormBase. These strains are usually available to order via stock centers in the USA and Japan, as a result of systematic genome-wide mutation projects and strains created in individual labs. Most of the older mutant strains were generated using either chemical (e.g., ethyl methane sulfate (EMS)) or transposon-mediated mutagenesis. Indeed, many of the mutants isolated in Brenner's original EMS screen [20] are still used today, often as the standard "reference" allele. However, CRISPR-based methods are now taking over as the preferred method for creating mutations in defined genes [29].

An alternative method for genome-wide screening is RNA interference (RNAi), which was pioneered in *C. elegans* [31]. The Nobel Prize-winning work of Fire and Mello showed that injection, soaking, or feeding of dsRNA solution leads to degradation of the corresponding RNA and an RNAi response in worms [32, 33]. Bacterial strains were created harboring plasmids encoding dsRNA, which the *C. elegans* subsequently feed upon. RNAi feeding is advantageous as it is extremely easy to perform and the bacterial strains created are permanently propagatable reagents. As a result, RNAi feeding libraries were constructed with the aim of covering all *C. elegans* genes to enable systematic genome-wide screening [34, 35]. The current RNAi bacterial feeding library consists of over 16,000 bacterial strains kept as glycerol stocks, equating to 87% of *C. elegans* genes. This has enabled many labs to perform genome-wide RNAi screens to identify genes involved in processes ranging from lifespan regulation to neurodegeneration [36, 37]. The high-throughput nature of RNAi screening and its applicability to 96-well plates make it particularly attractive for drug screening approaches.

Mutations in various *C. elegans* genes have been shown to cause spontaneous or increased susceptibility to seizure-like phenotypes [11, 12, 19]. All of these worm genes have human homologues that have been implicated in epilepsy [11, 12, 38, 39]. For a number of these genes, work in *C. elegans* was vital in both discovering and functionally characterizing the encoded proteins [11, 38, 39]. The first study of *C. elegans* seizure-like behavior came from the Caldwell Lab and centered on the epilepsy-associated disorder, lissencephaly [11]. Lissencephaly has been linked to mutations in the human *LIS1* gene. Williams et al. [11] found that mutations in the worm *LIS1* homologue (*lis-1*) resulted in characteristic head-bobbing convulsions in response to PTZ treatment. Similar convulsions were exhibited by worms harboring mutations in components of the GABAergic presynaptic signalling machinery. Through further investigation using RNAi knockdown of dynein and mutation of kinesin, it was concluded that the effects of *lis-1* on GABAergic presynaptic signalling were due to defective transport via neuronal cytoskeletal motor proteins [11].

A second example involved a study of the subunit composition of the cholinergic receptors expressed in motor neurons, where a gain of function mutant was identified [12]. The mutation mapped to the ACR-2 subunit pore domain, identical to a mutation in the epilepsy-associated disorder, myasthenia gravis. In *C. elegans* this resulted in spontaneous muscle convulsions through hyperactivation of cholinergic neurons and inactivation of GABAergic motor neurons. In follow-up investigations, a forward genetic screen revealed that the TRPM channel GTL-2 and its regulation of ion homeostasis were crucial for this mutant to develop its physiological and behavioral effects [15] and revealed that neuropeptides modulated this homeostasis [16]. These studies highlighted novel future targets for therapeutic screening and demonstrate how *C. elegans* can be appropriated for epilepsy research.

The nematode also offers several practical advantages for constructing high-throughput drug screens. *C. elegans* develop rapidly from egg to adulthood (defined as capable of egg laying) in 3 days. This means that compound efficacy can be established rapidly on both the developing and adult nervous system. Each worm also produces around 300 offspring by self-fertilization, and large populations can be maintained on 60 mm plates. Therefore turnaround for preparing viable adults for testing is short, and animal husbandry is both time- and cost-effective. The small size of the worm (around 1 mm in length as adults) also lends itself well to upscaling assays to multi-well plate formats, allowing larger-scale treatment screens without taking up excessive lab space [2, 10]. *C. elegans* is also the only animal model that can be frozen indefinitely and thawed for later use in a viable state.

Despite these several advantages, it is important to note limitations of this model. Firstly, the simplicity of *C. elegans* means that it is often difficult to directly relate research findings to complex organisms like humans. Hence, validation of promising findings in a more evolutionarily appropriate model may be necessary. Secondly one of the largest groups of genes associated with genetic epilepsy is those of the ligand- and voltage-gated ion channels. Many of these, including $GABA_A$ receptors, nicotinic acetylcholine receptors, and voltage-gated calcium and potassium channels, have similar functional properties and high degrees of conservation in *C. elegans* [40–42]. However, voltage-gated sodium channels are not present in the animal [42], and this should be taken into consideration during experimental design. *C. elegans* instead uses graded potentials, and calcium mediated spiking to propagate neuronal signals [43, 44]. However, as a novel drug discovery tool, this could be advantageous, as voltage-gated sodium channels are the main therapeutic target of current AED medication. Often hits in compound screens are found to affect voltage-gated sodium channel function, thereby saturating the drug discovery process. Using *C. elegans* for screening should select away from such compounds and therefore be more likely to identify AEDs with novel molecular targets and mechanisms of action. This, in turn, may help to find treatments for the approximately 30% of people with epilepsy who are resistant to current medication.

Established paradigms for testing potential new epilepsy therapies in rodents focus initially on the test compound's anti-seizure capabilities in acute seizure models [1, 45]. These include seizures induced electrically via corneal electrode stimulation, or induced chemically through exposure to convulsants such as pentylenetetrazol (PTZ). Thus, it is potentially important to distinguish between tests that show compounds are anti-seizure and which are antiepileptic. The *C. elegans* models that have been used to date for compound testing, including our own assay that this chapter describes, are all examples of anti-seizure screening [13, 14, 19, 46]. The PTZ seizure test in rodents uses a convulsive dose that induces a 5 s clonic seizure in 97% of the animals tested through subcutaneous injection [45]. Compounds can then be tested for their ability to protect against this seizure phenotype. The mechanism of action of PTZ remains uncertain. It is often assumed to be entirely due to actions on the ionotropic $GABA_A$ receptor; however, it has also been shown to have activity on potassium Kv1.1 channels as well as reducing calcium channel ion selectivity [47, 48]. A *C. elegans* model also provides evidence of some off-target activity, as PTZ was still seen to cause convulsions in a strong loss of function *unc-25* mutant, the only GABA synthesis gene known in the animal [11, 14].

As previously mentioned, the first report of seizure-like activity in *C. elegans* was in a study where PTZ was shown to induce head-bobbing convulsions in worms harboring mutations in *lis-1* and genes involved in GABAergic signalling [11]. In this and subsequent studies, PTZ was incorporated into the agar-based nematode growth medium. For drug screens this has a number of disadvantages, as creating the plates is labor-intensive, tedious, and incompatible with modern high-throughput procedures. Our group have adapted this assay, incorporating potential AEDs and PTZ into Dent's solution (used for observing basic behaviors). This method overcomes these issues and begins to make AED screening in *C. elegans* a potential high-throughput alternative. Using our method negates the need for drug plates, replacing them with PTZ formulation in liquid solution. This creates more stable solutions of PTZ, increases throughput, and reduces cost. Secondly PTZ exposure in liquid halves the exposure time needed for optimal induction of convulsions. Finally this approach is unaffected by DMSO concentrations up to 1%. This facilitates the use of compound libraries that are normally prepared as DMSO stocks which can be diluted and prepared in micro-titer formats.

2 Materials

2.1 Animal Handling

Worm manipulations and movement were conducted using platinum wire (Agar Scientific E404-2 platinum wire 0.2 mm diameter) attached to a worm pick handle. The wire is flattened using a coin to form a scoop which can then be used to move the *C. elegans*. The wire is sterilized between each pick over the flame of an ethanol burner with sterile technique followed throughout.

2.2 Microscope and Stage

To record for phenotypic analysis (either by eye or via automated analysis software), worms were recorded using a Zeiss Stemi 2000-C stereomicroscope (Carl Zeiss Limited, Cambridge, USA) which was mounted onto a Prior OptiScanIITM motorized microscope stage (Prior Scientific Inc., Massachusetts, USA). Worm Tracker (software version 2.0.3.1; http://www.mrc-lmb.cam.ac.uk/wormtracker) was used to record videos of seizure-like activity. Note that it is not essential to use these specific pieces of equipment; any stereomicroscope and video recording equipment/software are compatible. Note also that for some phenotypes, such as head-bobbing convulsions in *unc-49* worms, the animals are partially paralyzed by PTZ and motion tracking isn't strictly necessary. For higher-quality videos during liquid-based assay approaches, it should also be noted that having an inverted agar plate underneath the animal being imaged provides better contrast and easier visual analysis.

2.3 Animal Husbandry

Prior to analysis, worms are maintained on OP50 *E. coli*-seeded nematode growth medium (NGM) at 20 °C. NGM consists of 1 mM $CaCl_2$, 1 mM $MgSO_4$, 25 mM KH_2PO_4, 5μg/mL cholesterol in w/v 2% agar, 0.25% peptone, and 0.3% NaCl. Approximately 10 mL of NGM is dispensed in each 60 mm plastic petri dish, and ~5 worms are sub-cultured following OP50 seeding. To maintain healthy stocks, worms should be sub-cultured onto new seeded NGM plates every 3–5 days to prevent infections and over-population of plates.

Worms are age-synchronized for each of the described assays. The simplest way in which to do this is via "bleaching" (two parts 8% commercial alkaline hypochlorite bleach and one part 5 M NaOH). Worms are washed from well-populated (but not starved) plates using sterilized M9 buffer (0.5 g NaCl, 0.3 g KH_2PO_4, 0.6 g Na_2HPO_4, 0.1 mL 1 M $MgSO_4$ in 100 mL dH_2O) into a 15 mL falcon tube. Worms are then pelleted and aspirated and the bleach mixture added. Vortexing the mixture every 2 min for 10 min releases the eggs from the adult worms. Eggs are pelleted again (1 min, $1300 \times g$) and deposited onto seeded plates. A Day 1 adult (capable of egg laying) is considered to be at approximately 3.5 days following this procedure for wild-type worms. Depending on the genetic background, this can vary, as some mutant strains develop slowly; but for the assays described herein, 3.5 days is generally applicable.

2.4 Behavioral Solutions Required

All worm observations are conducted in Dent's solution (140 mM NaCl, 1 mM $MgCl_2$, 3 mM $CaCl_2$, 6 mM KCl, 10 mM HEPES, pH 7.4) with 0.1% bovine serum albumin (BSA) added alongside any treatment compounds on the day of testing. BSA prevents the worm's cuticle adhering to the plasticware used for observation. DMSO, ethanol, and beta-cyclodextrin solvents have also been added for compounds with limited aqueous solubility. DMSO can be added up to 1%; beyond this point toxic effects are seen. Ethanol should be used with appropriate controls due to it having inherent anticonvulsant properties. Sterile technique should be followed at all times with solutions and equipment sterilized.

3 Methods

PTZ had been used as a model of clonic-tonic seizure in mouse models and as such an identifier of seizure spread in rodents. It was first identified to have effects in worms with disrupted GABAergic signalling during a study of the effect of *lis-1* associated with lissencephaly [11]. During this research it was found that a characteristic head-bob convulsive phenotype could be invoked in the presence of PTZ, alongside a mild paralysis phenotype potentially from tonic contraction of the posterior half of the worm. These initial studies of convulsions relied on PTZ being dissolved into NGM plates with worms being placed onto these plates and measured for convulsions following 30 min. We have developed this further into a liquid-based assay more amenable to high-throughput applications. Our approach reduces exposure time necessary to induce seizures, and is more applicable to compound libraries, as these are typically diluted in DMSO.

Of all the worm strains tested, we have found the most robust and reproducible seizure-like behavior is produced from the *unc-49 (e407)* worm strain [19]. The *unc-49* gene encodes an ionotropic $GABA_A$ receptor that exhibits homology with human $GABA_A$ receptor subunits. It is unusual in that the single *unc-49* gene encodes all subunits required for the functioning oligomeric channel [49]. UNC-49 protein is mainly localized to muscle cells at neuromuscular junctions, but six other ligand-gated $GABA_A$ channels are expressed in worms. The *e407* allele we use routinely in the assay introduces a stop codon and has been demonstrated to be a genetic null mutation [49, 50]. We have also found using this liquid-based assay that PTZ induces convulsive phenotypes in *unc-25* mutants, *unc-43* mutants, and a humanized *STXBP1* epilepsy model. Note that spontaneous convulsions are rarely observed in any of these mutants: PTZ treatment is essential to induce seizure-like activity. In contrast, wild-type (Bristol N2) worms do not show seizure-like behavior in response to PTZ, although they do display a slow PTZ-induced paralysis phenotype at higher concentrations.

We have shown that age of worms produces no significant difference between Day 1 and Day 5 adult worms [19]. However, age synchronization should be considered to reduce variability in the assay. Age synchronization of worms can be conducted in numerous ways, including timed egg laying and bleaching. We typically perform age synchronization via bleaching, as this ensures clean stock plates from which to perform behavioral phenotyping. Semi-starved plates of C. elegans are washed using sterile M9 solution with an equal measure of bleach and 1 M NaOH (two parts 8% commercial alkaline hypochlorite bleach and one part 5 M NaOH). This dissolves the adult worms but spares the eggs. This solution is then pipetted onto fresh plates providing a relatively homogeneous aged population. This should be taken into account when planning as from this point in 3 days, they will be considered Day 1 adults (capable of egg laying).

As a low-/medium-throughput assay, this can be conducted on either glass slides or empty petri dishes. It should be noted that when conducting any C. elegans phenotypic assay on plasticware, bovine serum albumin (BSA) should be added in order to avoid the worms adhering to the plastic. The assay of seizure-like activity can be divided into three steps: (1) pre-incubation; (2) PTZ incubation; and (3) observation (Fig. 1). Seizure-like activity is defined here as the repetitive extension and contraction of the head/pharyngeal region of the worm along the anterior-posterior axis (Fig. 2a).

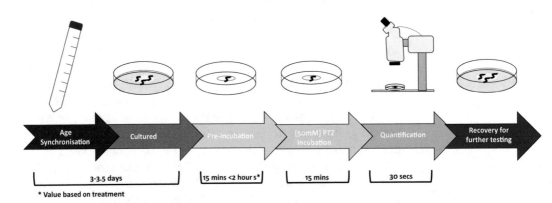

Fig. 1 Overview workflow of the PTZ assay. Worms are first cleaned/age-synchronized and upon reaching adulthood subsequently undergo pre-incubation, PTZ incubation, and observation. Pre-incubation time with compounds of interest varies due to differences in the compounds' ability to penetrate the worm cuticle

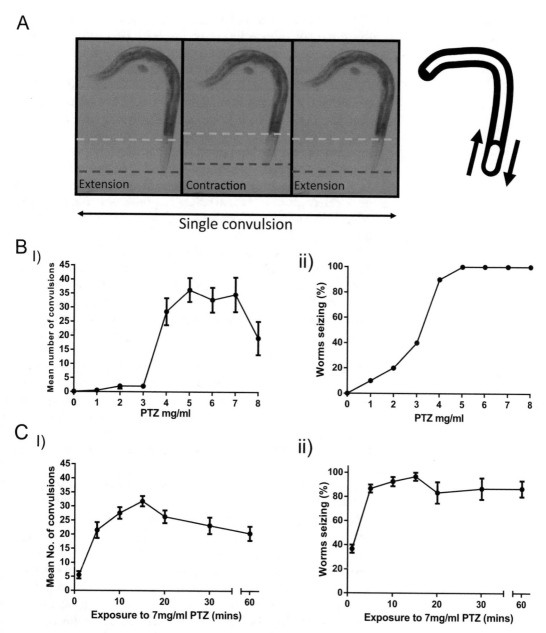

Fig. 2 PTZ exposure at 7 mg/mL for 15 min causes optimum seizure-like activity in *C. elegans unc-49(e407)* mutants. (**a**) Still frame images and cartoon showing the convulsion phenotype defined as extension and contraction of the head/pharyngeal region along the anterior-posterior axis. The green dashed line represents a reference for position of the pharynx, and red line represents tip of the head. (**b**) Convulsion concentration response for PTZ between 1 mg/mL and 8 mg/mL ($n = 10$). (i) Total convulsions over a 30 s interval following 15 min PTZ incubation ($n = 10$). (ii) Worms displaying a seizure phenotype defined as three consecutive convulsions in rapid succession. (**c**) Time-response experiments against 7 mg/mL PTZ. (i) Total number of convulsions over the 30 s measurement interval ($n = 30$). (ii) Worms displaying a seizure phenotype defined as three consecutive convulsions in rapid succession ($n = 30$)

3.1 Step-by-Step Procedure

1. *Pre-incubation*. Pick individual worms into 60 μL droplets of the pre-incubation solution of interest. This can be control solutions containing Dent's solution and solvent vehicle, or Dent's solution containing an experimental compound. *C. elegans* has an outer cuticle which can dramatically reduce penetrance and bioaccumulation of many compounds [51], thus causing a delay in compound absorption and effect. As such we recommend the following analysis when designing experiments for testing potential AEDs. First, assess the compound's predicted permeability and bioaccumulation in *C. elegans* using an in silico published algorithm [51]. Second, perform empirical concentration response and time to peak effect experiments to establish optimal drug treatment paradigms for each compound. Peak effect times vary between 15 min and 2 h, and optimal concentrations range between 150 μM and 15 mM for individual AEDs, based on our experience. Note that these concentrations appear high because they represent those in the external bathing medium. The barrier formed by the worm cuticle typically means that the internal drug concentration is orders of magnitude lower than that in the external media [51]. For example, only around 1% of the AED ethosuximide that is applied externally actually penetrates inside the worm, as measured by NMR [46]. DMSO is routinely used as a vehicle in this assay at up to 1% concentration, but ethanol and beta-cyclodextrin can also be used as formulation vehicles with negligible adverse effects at the concentrations needed. Above 1% DMSO, toxic effects are seen, which can interfere with the convulsive phenotype and health of the animal. Note that, due to the small size of the droplets, you should keep pre-incubating worms covered to prevent evaporation altering the concentrations of compounds.

2. *PTZ incubation*. Transfer worms to new droplets containing both PTZ and the test AED compound (at the same concentration used for pre-incubation). For the unc-49(e407) *C. elegans* strain that we routinely use, the optimum PTZ response is produced at 7 mg/mL, which is approximately 50 mM (Fig. 2b). As already mentioned, although this concentration appears high, the worm cuticle lowers the penetrance of compounds, and so the concentration of PTZ inside the worm is likely much lower (although we have not directly measured this). Although some worms convulse in the first few minutes, we have found the optimum effect of PTZ to be after 15 min incubation with 50 mM PTZ (Fig. 2c).

3. *Observation*. Worms will now show convulsion phenotypes if no AED pre-incubation is present; *unc-49(e407)* worms will display anterior convulsions alongside tonic paralysis of the

posterior of the worm (*see* **Note 1** and supplementary movie). Once measurements have been taken, recover worms from the solution if necessary to use in further analysis (also refer to **Notes 2** and **3**).

3.2 Data Acquisition and Interpretation

Data should ideally be acquired using video microscopy. This enables assessment of convulsions by independent observers, who can be blinded if desired to reduce bias. Two approaches can be taken to score convulsions: either quantitatively or qualitatively. To score quantitatively, count the number of convulsions over a pre-defined time (we normally use 30 s). To score qualitatively, use the following criteria for a "seizure-like" phenotype: three or more rapid convulsions over the measurement period. We typically use groups of ten worms with three biological repeat experiments for each condition we are testing. Perform statistical analysis as appropriate.

4   Notes

1. *C. elegans unc-49(e407)* convulsions are subtle and can be hard to distinguish from standard head foraging movements. It can help to identify the terminal bulb of the pharynx and watch for its contraction into the body, an observation that is more exaggerated in a convulsion in comparison to other head movements.

2. Sometimes particular chemical treatment regimens can be toxic, or damage can occur during the manipulation of the *C. elegans* leading to the death of the worm. If this begins to be a concern, perform simple mechanosensation response tests in these worms; tap either the head or tail of the concerned worm three times, and check for responses.

3. As an optional addition, transfer the affected worm to a droplet containing only Dent's solution and watch for recovery.

5   Conclusion

PTZ and other means of acute seizure induction continue to provide invaluable information when conducting first-pass screens of lead compounds in rodents. This concerns both neurological activity of new compounds and the ability of a compound to prevent spread of synchronous activity in the brain. Incorporation of the nematode *C. elegans* provides a more ethical and cost-efficient addition to these initial screens. The described protocol provides the basis for potential high-throughput adaptions and a companion assay to traditional approaches. This approach can allow testing of small compound libraries, or of a series of chemical variations of active hit compounds for structure-activity relationship studies, without adding additional mammalian costs.

In our group, we have used this simple behavioral assay both in genetic and pharmacological projects. For example, we have used our method of PTZ exposure in the characterization of epilepsy-associated genetic variants in the human *STXBP1* gene [52]. A humanized strain was created via introduction of the *STXBP1* gene into an *unc-18* (*C. elegans STXBP1* orthologue) null mutant worm strain. PTZ-induced convulsions were observed in several worm strains expressing *STXBP1* mis-sense mutations found in epilepsy patients, but not in worms expressing wild-type *STXBP1*. Hence, the PTZ assay can be used to rapidly assess the pathogenicity of human genetic variants associated with epilepsy.

In other studies we have utilized the worm's facile genetics and the PTZ assay to characterize existing and new anticonvulsant compounds. In our most recent methods paper, we discuss the application of this assay to verify the mechanism of action of the AED, ethosuximide [19]. It is commonly assumed that ethosuximide's molecular target is the T-type calcium channel, although various other targets have also been suggested [53]. However, using a null mutation in the worm *cca-1* gene (orthologue of the T-type calcium channel), we demonstrated that *cca-1* does not affect ethosuximide's anticonvulsant action [19]. In addition to its action as an AED, ethosuximide also protects against neurodegeneration in worm, cell culture, and rodent models [54–56]. However, its neuroprotective effect requires high concentrations, so we set out to identify ethosuximide derivatives with higher potency using a combination of chemoinformatics and initial compound screening using the worm PTZ assay [46]. This identified the compound alpha-methyl-alpha-phenylsuccinimide as being 100-fold more potent than ethosuximide in protecting against neurodegeneration.

We have also demonstrated the translational potential of our assay. In an ongoing project, we used *D. rerio* (zebrafish) for initial high-throughput screening, then employed secondary testing and chemical genetic characterization in worms, and finally hit compound validation in mouse models and human tissue (Jones, A., Barker-Haliski, M., Ilie, A.S., Herd, M.B. Baxendale, S., Holdsworth, C.J., Ashton, J.P., Placzek, M., Jayasekera, B.A.P., Cowie, C.J.A., Lambert, J.J., Trevelyan, A.J., White, H.S., Marson, A.G., Cunliffe, V.T., Sills, G.J. and Morgan, A. (2020) A multiorganism pipeline for antiseizure drug discovery: Identification of chlorothymol as a novel γ-aminobutyric acidergic anticonvulsant. Epilepsia 61, 2106–2118.). This identified a novel anticonvulsant compound in both *D. rerio* and *C. elegans*, suggesting an evolutionarily conserved target between both animals. Subsequently, we discovered using a worm genetic screen that the compound's molecular target is a GABA$_A$ receptor. The drug's anticonvulsant activity was conserved in mice utilizing a number of standard seizure models, being particularly effective against electrically induced partial seizures. Finally, the GABAergic mechanism of action was replicated and measured electrophysiologically in human and mouse neuronal tissue. Thus, using PTZ models in fish and worms has the potential to identify novel anticonvulsants with evolutionarily conserved mechanisms of action and translational potential.

One drawback of the technique described here is that it relies on the manual picking of worms into droplets and the visual scoring of convulsions by a human observer. This makes the assay labor-intensive and subject to variability. The current assay could therefore be further improved and made suitable for high-throughput AED screening by using established robotic systems that are already available for nematodes. These techniques are readily adaptable for micro-titer plate configurations and can automate the culturing and handling of the worms [57–59]. Furthermore, using automated image analysis techniques and phenotype tracking software would obviate the need for human observers and greatly speed up the assay. Given that such automated imaging and tracking technologies have already been developed for other phenotypes in *C. elegans* [60], this is likely achievable for the head-bobbing phenotype.

In summary, we have detailed our PTZ convulsion assay, which can be incorporated into high-throughput AED screening with the addition of appropriate automated hardware/software. We have validated this approach with clinically approved AEDs and demonstrated its potential for identifying new anticonvulsants and revealing mechanisms of AED action. Addition of *C. elegans* to current studies provides a cheap, ethical animal model to complement traditional rodent approaches for AED screening.

Acknowledgments

AJ was supported by a PhD studentship funded by the MRC. Strains used in this work were provided by the *Caenorhabditis* Genetics Center, which is funded by the NIH National Center for Research Resources (NCRR).

References

1. Loscher W (2017) Animal models of seizures and epilepsy: past, present, and future role for the discovery of antiseizure drugs. Neurochem Res 42:1873–1888
2. Cunliffe VT, Baines RA, Giachello CN, Lin WH, Morgan A, Reuber M, Russell C, Walker MC, Williams RS (2015) Epilepsy research methods update: understanding the causes of epileptic seizures and identifying new treatments using non-mammalian model organisms. Seizure 24:44–51

3. Barton ME, Klein BD, Wolf HH, White HS (2001) Pharmacological characterization of the 6 Hz psychomotor seizure model of partial epilepsy. Epilepsy Res 47:217–227

4. White HS, Johnson M, Wolf HH, Kupferberg HJ (1995) The early identification of anticonvulsant activity: role of the maximal electroshock and subcutaneous pentylenetetrazol seizure models. Ital J Neurol Sci 16:73–77

5. Rowley NM, White HS (2010) Comparative anticonvulsant efficacy in the corneal kindled mouse model of partial epilepsy: correlation with other seizure and epilepsy models. Epilepsy Res 92:163–169

6. Barker-Haliski M, Harte-Hargrove LC, Ravizza T, Smolders I, Xiao B, Brandt C, Loscher W (2018) A companion to the preclinical common data elements for pharmacologic studies in animal models of seizures and epilepsy. A Report of the TASK3 Pharmacology Working Group of the ILAE/AES Joint Translational Task Force. Epilepsia Open 3:53–68

7. Baraban SC (2007) Emerging epilepsy models: insights from mice, flies, worms and fish. Curr Opin Neurol 20:164–168

8. Baines RA, Giachello CNG, Lin W-H (2017) Chapter 24 – Drosophila. In: Pitkänen A, Buckmaster PS, Galanopoulou AS, Moshé SL (eds) Models of seizures and epilepsy, 2nd edn. Academic, London, pp 345–358

9. Copmans D, Siekierska A, de Witte PAM (2017) Chapter 26 – zebrafish models of epilepsy and epileptic seizures. In: Pitkänen A, Buckmaster PS, Galanopoulou AS, Moshé SL (eds) Models of seizures and epilepsy, 2nd edn. Academic, London, pp 369–384

10. Chen X, Barclay JW, Burgoyne RD, Morgan A (2015) Using C. elegans to discover therapeutic compounds for ageing-associated neurodegenerative diseases. Chem Cent J 9:65

11. Williams SN, Locke CJ, Braden AL, Caldwell KA, Caldwell GA (2004) Epileptic-like convulsions associated with LIS-1 in the cytoskeletal control of neurotransmitter signaling in Caenorhabditis elegans. Hum Mol Genet 13:2043–2059

12. Jospin M, Qi YB, Stawicki TM, Boulin T, Schuske KR, Horvitz HR, Bessereau JL, Jorgensen EM, Jin Y (2009) A neuronal acetylcholine receptor regulates the balance of muscle excitation and inhibition in Caenorhabditis elegans. PLoS Biol 7:e1000265

13. Pandey R, Gupta S, Tandon S, Wolkenhauer O, Vera J, Gupta SK (2010) Baccoside A suppresses epileptic-like seizure/convulsion in Caenorhabditis elegans. Seizure 19:439–442

14. Risley MG, Kelly SP, Jia K, Grill B, Dawson-Scully K (2016) Modulating behavior in C. elegans using electroshock and antiepileptic drugs. PLoS One 11:e0163786

15. Stawicki TM, Zhou K, Yochem J, Chen L, Jin Y (2011) TRPM channels modulate epileptic-like convulsions via systemic ion homeostasis. Curr Biol 21:883–888

16. Stawicki TM, Takayanagi-Kiya S, Zhou K, Jin Y (2013) Neuropeptides function in a homeostatic manner to modulate excitation-inhibition imbalance in C. elegans. PLoS Genet 9:e1003472

17. Takayanagi-Kiya S, Jin Y (2017) Chapter 23 – nematode C. elegans: genetic dissection of pathways regulating seizure and epileptic-like behaviors A2 – Pitkänen, Asla. In: Buckmaster PS, Galanopoulou AS, Moshé SL (eds) Models of seizures and epilepsy, 2nd edn. Academic, London, pp 327–344

18. Locke CJ, Caldwell KA, Caldwell GA (2009) The nematode, Caenorhabditis elegans, as an emerging model for investigating epilepsy. In: Baraban SC (ed) Animal models of epilepsy: methods and innovations. Humana Press, Totowa, pp 1–25

19. Wong SQ, Jones A, Dodd S, Grimes D, Barclay JW, Marson AG, Cunliffe VT, Burgoyne RD, Sills GJ, Morgan A (2018) A Caenorhabditis elegans assay of seizure-like activity optimised for identifying antiepileptic drugs and their mechanisms of action. J Neurosci Methods 309:132–142

20. Brenner S (1974) The genetics of *Caenorhabditis elegans*. Genetics 77:71–94

21. Sulston JE, Horvitz HR (1977) Post-embryonic cell lineages of the nematode, Caenorhabditis elegans. Dev Biol 56:110–156

22. Sulston JE, Schierenberg E, White JG, Thomson JN (1983) The embryonic cell lineage of the nematode Caenorhabditis elegans. Dev Biol 100:64–119

23. White JG, Southgate E, Thomson JN, Brenner S (1986) The structure of the nervous system of the nematode Caenorhabditis elegans. Philos Trans R Soc Lond Ser B Biol Sci 314:1–340

24. Barclay JW, Morgan A, Burgoyne RD (2012) Neurotransmitter release mechanisms studied in Caenorhabditis elegans. Cell Calcium 52:289–295

25. International League Against Epilepsy Consortium on Complex, E (2018) Genome-wide mega-analysis identifies 16 loci and highlights diverse biological mechanisms in the common epilepsies. Nat Commun 9:5269

26. Epi4K Consortium (2013) De novo mutations in epileptic encephalopathies. Nature 501:217

27. Noebels J (2015) Pathway-driven discovery of epilepsy genes. Nat Neurosci 18:344–350

28. Kim W, Underwood RS, Greenwald I, Shaye DD (2018) OrthoList 2: a new comparative genomic analysis of human and Caenorhabditis elegans genes. Genetics 210:445–461

29. Dickinson DJ, Goldstein B (2016) CRISPR-based methods for Caenorhabditis elegans genome engineering. Genetics 202:885–901

30. Au V, Li-Leger E, Raymant G, Flibotte S, Chen G, Martin K, Fernando L, Doell C, Rosell FI, Wang S, Edgley ML, Rougvie AE, Hutter H, Moerman DG (2019) CRISPR/Cas9 methodology for the generation of knockout deletions in Caenorhabditis elegans. G3 9:135–144

31. Conte D Jr, MacNeil LT, Walhout AJ, Mello CC (2015) RNA interference in Caenorhabditis elegans. Curr Protoc Mol Biol 109:26.3.1–26.3.30

32. Fire A, Xu S, Montgomery MK, Kostas SA, Driver SE, Mello CC (1998) Potent and specific genetic interference by double-stranded RNA in Caenorhabditis elegans. Nature 391:806–811

33. Timmons L, Fire A (1998) Specific interference by ingested dsRNA. Nature 395:854

34. Kamath RS, Fraser AG, Dong Y, Poulin G, Durbin R, Gotta M, Kanapin A, Le Bot N, Moreno S, Sohrmann M, Welchman DP, Zipperlen P, Ahringer J (2003) Systematic functional analysis of the Caenorhabditis elegans genome using RNAi. Nature 421:231–237

35. Rual JF, Ceron J, Koreth J, Hao T, Nicot AS, Hirozane-Kishikawa T, Vandenhaute J, Orkin SH, Hill DE, van den Heuvel S, Vidal M (2004) Toward improving Caenorhabditis elegans phenome mapping with an ORFeome-based RNAi library. Genome Res 14:2162–2168

36. Murphy CT, McCarroll SA, Bargmann CI, Fraser A, Kamath RS, Ahringer J, Li H, Kenyon C (2003) Genes that act downstream of DAF-16 to influence the lifespan of Caenorhabditis elegans. Nature 424:277–283

37. Nollen EA, Garcia SM, van Haaften G, Kim S, Chavez A, Morimoto RI, Plasterk RH (2004) Genome-wide RNA interference screen identifies previously undescribed regulators of polyglutamine aggregation. Proc Natl Acad Sci U S A 101:6403–6408

38. Hosono R, Hekimi S, Kamiya Y, Sassa T, Murakami S, Nishiwaki K, Miwa J, Taketo A, Kodaira KI (1992) The unc-18 gene encodes a novel protein affecting the kinetics of acetylcholine metabolism in the nematode Caenorhabditis elegans. J Neurochem 58:1517–1525

39. Guiberson NGL, Pineda A, Abramov D, Kharel P, Carnazza KE, Wragg RT, Dittman JS, Burre J (2018) Mechanism-based rescue of Munc18-1 dysfunction in varied encephalopathies by chemical chaperones. Nat Commun 9:3986

40. Caylor RC, Jin Y, Ackley BD (2013) The Caenorhabditis elegans voltage-gated calcium channel subunits UNC-2 and UNC-36 and the calcium-dependent kinase UNC-43/CaMKII regulate neuromuscular junction morphology. Neural Dev 8:10

41. Steger KA, Shtonda BB, Thacker C, Snutch TP, Avery L (2005) The C. elegans T-type calcium channel CCA-1 boosts neuromuscular transmission. J Exp Biol 208:2191–2203

42. Bargmann CI (1998) Neurobiology of the Caenorhabditis elegans genome. Science 282:2028–2033

43. Gao S, Guan SA, Fouad AD, Meng J, Kawano T, Huang YC, Li Y, Alcaire S, Hung W, Lu Y, Qi YB, Jin Y, Alkema M, Fang-Yen C, Zhen M (2018) Excitatory motor neurons are local oscillators for backward locomotion. elife 7:29915

44. Gao S, Zhen M (2011) Action potentials drive body wall muscle contractions in Caenorhabditis elegans. Proc Natl Acad Sci U S A 108:2557–2562

45. Löscher W (2011) Critical review of current animal models of seizures and epilepsy used in the discovery and development of new antiepileptic drugs. Seizure 20:359–368

46. Wong SQ, Pontifex MG, Phelan MM, Pidathala C, Kraemer BC, Barclay JW, Berry NG, O'Neill PM, Burgoyne RD, Morgan A (2018) Alpha-methyl-alpha-phenylsuccinimide ameliorates neurodegeneration in a C. elegans model of TDP-43 proteinopathy. Neurobiol Dis 118:40–54

47. Madeja M, Musshoff U, Lorra C, Pongs O, Speckmann EJ (1996) Mechanism of action of the epileptogenic drug pentylenetetrazol on a cloned neuronal potassium channel. Brain Res 722:59–70

48. Papp A, Feher O, Erdelyi L (1987) The ionic mechanism of the pentylenetetrazol convulsions. Acta Biol Hung 38:349–361

49. Bamber BA, Beg AA, Twyman RE, Jorgensen EM (1999) The Caenorhabditis elegans unc-49 locus encodes multiple subunits of a heteromultimeric GABA receptor. J Neurosci 19:5348–5359

50. Jin Y, Jorgensen E, Hartwieg E, Horvitz HR (1999) The Caenorhabditis elegans gene unc-25 encodes glutamic acid decarboxylase and is required for synaptic transmission but not synaptic development. J Neurosci 19:539–548

51. Burns AR, Wallace IM, Wildenhain J, Tyers M, Giaever G, Bader GD, Nislow C, Cutler SR, Roy PJ (2010) A predictive model for drug bioaccumulation and bioactivity in Caenorhabditis elegans. Nat Chem Biol 6:549–557

52. Zhu B, Mak JCH, Morris AP, Marson AG, Barclay JW, Sills GJ, Morgan A (2020) Functional analysis of epilepsy-associated variants in STXBP1/Munc18-1 using humanized Caenorhabditis elegans. Epilepsia 61:810

53. Crunelli V, Leresche N (2002) Block of thalamic T-type Ca(2+) channels by ethosuximide is not the whole story. Epilepsy Curr/Am Epilepsy Soc 2:53–56

54. Chen X, McCue HV, Wong SQ, Kashyap SS, Kraemer BC, Barclay JW, Burgoyne RD, Morgan A (2015) Ethosuximide ameliorates neurodegenerative disease phenotypes by modulating DAF-16/FOXO target gene expression. Mol Neurodegener 10:51

55. Tauffenberger A, Julien C, Parker JA (2013) Evaluation of longevity enhancing compounds against transactive response DNA-binding protein-43 neuronal toxicity. Neurobiol Aging 34:2175–2182

56. Tiwari SK, Seth B, Agarwal S, Yadav A, Karmakar M, Gupta SK, Choubey V, Sharma A, Chaturvedi RK (2015) Ethosuximide induces hippocampal neurogenesis and reverses cognitive deficits in an amyloid-beta toxin-induced Alzheimer rat model via the phosphatidylinositol 3-kinase (PI3K)/Akt/Wnt/beta-catenin pathway. J Biol Chem 290:28540–28558

57. Bazopoulou D, Chaudhury AR, Pantazis A, Chronis N (2017) An automated compound screening for anti-aging effects on the function of C. elegans sensory neurons. Sci Rep 7:9403

58. Boyd WA, McBride SJ, Rice JR, Snyder DW, Freedman JH (2010) A high-throughput method for assessing chemical toxicity using a Caenorhabditis elegans reproduction assay. Toxicol Appl Pharmacol 245:153–159

59. Xian B, Shen J, Chen W, Sun N, Qiao N, Jiang D, Yu T, Men Y, Han Z, Pang Y, Kaeberlein M, Huang Y, Han Jing-Dong J (2013) WormFarm: a quantitative control and measurement device toward automated Caenorhabditis elegans aging analysis. Aging Cell 12:398–409

60. Buckingham SD, Partridge FA, Sattelle DB (2014) Automated, high-throughput, motility analysis in Caenorhabditis elegans and parasitic nematodes: applications in the search for new anthelmintics. Int J Parasitol Drugs Drug Resist 4:226–232

INDEX

Divya Vohora (ed.), *Experimental and Translational Methods to Screen Drugs Effective Against Seizures and Epilepsy*,
Neuromethods, vol. 167, https://doi.org/10.1007/978-1-0716-1254-5,
© Springer Science+Business Media, LLC, part of Springer Nature 2021

Printed in the United States
by Baker & Taylor Publisher Services